THE ULTIMATE BAOFENG RADIO BIBLE

The Complete Guerrilla's Tactical Guide To Mastering The Baofeng Radio For Emergency Preparedness, Natural Disaster,War Scenarios, Safety And Survival In Every Crisis Situation | Top Techniques For Safety And Communication

SAGE VANGUARD

Thank You for Choosing This Book!

YOU'VE MADE THE RIGHT CHOICE

First and foremost, I want to extend my heartfelt gratitude for choosing **The Ultimate baofeng radio bible !** Your decision to embark on this journey toward mastering tactical communication and preparedness means the world.

As a token of gratitude, I've included exclusive bonuses to further enhance your experience:

Bonus #1: Scan the QR code below that leads to a file containing a comprehensive Trigam Encoding and Decoding List. This resource will amplify your understanding and usage of essential codes for secure communication.

Bonus #2: Don't miss the dedicated Bonus Section within the book, curated with additional insights, strategies, and practical tips to amplify your skills in emergency preparedness and crisis communication.

Your commitment to readiness and preparedness inspires me. Thank you for joining me on this important journey.

Stay safe, stay prepared!

Sage Vanguard

ABOUT THE AUTHOR

Sage Vanguard is a seasoned tactical communication expert with over two decades of experience in emergency preparedness and survival strategies. A former military communications specialist, Sage's expertise lies in empowering individuals to master the use of Baofeng Radios for guerrilla communication in crisis situations. His commitment to safety and preparedness has made him a trusted authority in the field, offering invaluable insights and comprehensive guidance for navigating emergencies, natural disasters, and conflicts. Sage's dedication to equipping individuals with the necessary tools and tactics to ensure safety in the most challenging scenarios shines through in his latest book, **THE ULTIMATE BAOFENG BIBLE.**

Table of Contents

War

Emergency

Natural Disaster

Survival

Communication

Every Crisis situation

We've Got You
Covered

INTRODUCTION TO BAOFENG RADIOS

Imagine a world where a simple handheld device becomes a lifeline in the midst of chaos. Enter the Baofeng radio — an unassuming yet powerful tool , known for its affordability and ease of use, operates on Very High Frequency (VHF) and Ultra High Frequency (UHF) bands with Frequency Modulation (FM). It's like a magical device that can do a lot, even if you're new to it.

This guide isn't just about pressing buttons; it's a manifesto for survival in the unknown. Whether you're a guerrilla, a dissident, or just someone interested in practical tools, the Baofeng is here to empower you with what you have on hand.

At its core, the Baofeng is more than just a radio; it's a guerrilla's best ally, the dissident's voice, and the activist's shield. In a world where conflict looms large, this manual unveils the secrets of turning this unassuming gadget into a lifeline amidst chaos.

Think of it like the Baofeng being your main communication tool, as important as any weapon. We'll keep things simple, focusing on using it effectively rather than getting into too much technical stuff.

No matter your skill level, this guide is for you. It's written with potential guerrilla users in mind, not like a complicated Amateur Radio book. But, you can still use some of the principles for your Ham radio hobby.

The Roles Of Communication

Now, let's talk about the Three Roles of Communications— Sustainment, Tactical, and Clandestine/Strategic. These are like different jobs your radio can do, each with its own unique considerations.

Sustainment Communications are like the friendly chats you have when your normal phones aren't working. It's all about public safety and survival, like creating networks in rural areas or after a natural disaster. It's not super secret, but we still need to be a bit careful.

Tactical Communications are like quick messages in the heat of action, such as coordinating between teams or passing info during a mission. They need to be short and secure, using codes and special antennas to limit who hears them.

Now, **Clandestine or Strategic Communications** are super strategic! They involve a lot of planning and the highest level of security. Imagine sending messages over a longer range, making sure no one intercepts them. It's like planning secret moves against a powerful opponent.

Baofeng Radio has some cool abilities! It can operate in different frequency ranges, like listening to international FM radio or even setting up a local pirate radio station for secret messages and friendly propaganda. It can also intercept military ground communications, acting like a secret listener to important info.

Remember, just because a transmission is encrypted doesn't mean it can't be intercepted. Even weird static or changes in signals can be like a secret warning. In Iraq, insurgents used analog TVs to sense changes in static and know when American troops were around. The Baofeng can play a similar role.

Baofeng Radio Frequency Ranges

67—108 MHz (RECEIVE ONLY)

This range is dedicated to receiving international FM radio broadcasts. It's not limited to music; you can use it to set up a local FM station for discreet communication.

136—174 MHz and 400—470 MHz (TRANSMIT/RECEIVE)

These are the main working frequencies of the Baofeng. It can both transmit and receive signals in these ranges. This versatility is handy for various communication needs, including tactical operations and intercepting military ground communications.

220—250 MHz (TRANSMIT/RECEIVE):

Expanding its reach, this range provides additional options for communication scenarios.

Decrypting Signals

Even if a transmission is encrypted, the Baofeng has a knack for picking up signals. This ability is comparable to how insurgents in Iraq used analog TVs to detect changes in static, serving as an early warning system.

Practical Applications:

Covert FM Radio Stations: Utilize the 67—108 MHz range to create hidden FM radio stations for transmitting resistance information.

Tactical Operations: Employ the 136—174 MHz and 400—470 MHz ranges for on-the-fly coordination during military-type operations.

Monitoring Military Communications: Intercept transmissions in the 136—174 MHz and 400—470 MHz ranges to stay informed about military activities.

- **US/NATO:** 30-88 MHz

- **CSTO/Russia/China/Venezuela:** 30-108 MHz

The Baofeng, with its versatile frequency coverage, aligns perfectly with the last half of both NATO and CSTO ground communication systems. This means it can intercept or monitor transmissions within these military frequencies.

Decrypting the Encryption:

Encryption doesn't guarantee immunity. Even if transmissions are encrypted, the Baofeng has the ability to intercept them. The presence of any kind of traffic, even if it's just unintelligible digital signals, in a particular area could serve as an early warning.

Practical Insight:

Early Warning System: Similar to insurgents in Iraq using analog TVs to detect changes in static, the Baofeng can play a role in providing early warnings about the presence of military activities.

The Baofeng's capability to tap into military frequencies makes it a strategic tool for staying informed about ground operations.

Frequency Insights:

Unlocking the Baofeng's Transmission Realm

The Baofeng radio operates as a dual band radio, showcasing its prowess in both VHF (Very High Frequency) and UHF (Ultra High Frequency) ranges across a broad spectrum of frequencies. This expansive range caters to diverse needs, from licensed frequencies requiring specific permissions to license-free frequencies.

Navigating this sea of frequencies demands a strategic approach. The key lies in avoiding predictability while acknowledging the full range of the radio's capabilities. Within these frequencies lie various services, each harboring different types of information and unique communication traffic.

Very High Frequency (VHF) - 136-174 MHz:

VHF proves its mettle in rural and hilly environments, demonstrating resilience against signal loss in dense vegetation. Boasting a longer range than its UHF counterpart and a higher power output, VHF shines in various services within its frequency range:

- **Multi-Use Radio Service (MURS):**

    A license-free oasis with five specific frequencies (channels), MURS provides a low-key, low-cost option for sustainment-level communications. Frequencies include 151.820, 151.880, 151.940, 154.570, and 154.600.

- **Business Band VHF:**

  A licensed service catering to commercial enterprises, transmitting both voice and data. Frequencies range from 151.505 MHz to 158.4075 MHz. Vigilance is crucial, considering potential shared usage among businesses and the presence of emergency service communications.

- **NOAA Weather:**

  NOAA employs seven frequencies, transmitting weather data and emergency relief information. Frequencies cover a range from 162.400 MHz to 162.550 MHz.

- **Marine Band:**

  Reserved for maritime vessels, operating in the frequency range of 155-160 MHz. Inland, it is also utilized by public service agencies, including law enforcement.

- **High Band VHF Amateur Radio (2 Meter):**

  Nicknamed for the physical length of one full wavelength (2 meters), this requires a license to operate in the 144-148 MHz frequency range. It plays a significant role in the amateur radio world.

Understanding these frequencies unlocks the Baofeng's potential for a myriad of applications. Whether monitoring weather data, engaging in tactical communications, or participating in amateur radio, the Baofeng stands ready.

Unveiling Cover For Action (CFA): A Cautionary Note

In the realm of Amateur Radio, where lawful operation demands a license and generates a trail of personal data, a unique cloak emerges—Cover For Action (CFA). While the license requirement safeguards communication integrity, it also serves as a disguise for activities that might otherwise raise suspicion under the guise of benign intentions.

An intriguing example involves Nellie Ohr, who obtained a Technician-level Ham radio license (KM4UDZ) before her involvement in leaking a dossier related to then-President Donald Trump. Allegedly directed by British freelance spy Christopher Steele, the Ohrs, situated in a region rich in radio monitoring equipment for counterintelligence, utilized the Ham radio license as a means to transmit files discreetly, avoiding potentially compromised channels.

In the event of discovery, their CFA could have been as simple as claiming the acquisition of equipment for a recreational hobby and the testing of Ham radio equipment in a digital mode. The effectiveness of a CFA often hinges on the investigator's lack of expertise in the field, coupled with the plausibility of the cover story.

Ultra High Frequency (UHF) - 400-470 MHz: Unlocking Urban Potential

The Baofeng's UHF range, spanning 400-470 MHz, excels in urban environments. However, in rural settings, it contends with higher signal loss in dense vegetation. UHF, known for penetrating buildings rather than reflecting signals off them, proves tactically advantageous by mitigating interception risks at the operational level.

For an adversary equipped with signals intelligence (SIGINT) interception tools, intercepting and gaining a line of bearing (LOB) on a low-power UHF signal in dense vegetation poses challenges. Tactical groups leveraging low power must acknowledge the trade-off: shorter physical ranges, often less than 1 kilometer. UHF services include:

- **Family Radio Service (FRS) and General Mobile Radio Service (GMRS):**

  Operating within the same frequency space, these services cater to common two-way radios found in sporting goods stores. FRS requires no license but has a low power output (2 watts), while GMRS, requiring a license, allows repeater use and up to 50 watts of power output.The frequency ranges in between 462.5625 MHz,462.7250 MHz.

- **UHF Business Band:**

  The counterpart to its VHF version, operating at frequencies like 464.500 MHz, 467.850 MHz, and others.

Navigating FRS/GMRS: Channelized with known published data, FRS/GMRS requires similar considerations for Communication Security (COMSEC) as MURS. Despite this, it offers a robust option for simple sustainment-level capability with minimal training.

70cm Amateur Radio: A Wave of Connectivity

Similar to its VHF counterpart, 70cm Amateur Radio derives its name from one full wavelength of a UHF signal in that band segment. Operating within the frequency range of 420-470 MHz, this band enjoys remarkable popularity in densely populated areas across the US and globally. Its versatility and widespread usage make it a go-to choice for radio enthusiasts navigating the airwaves.

220-250 MHz: The Enigmatic 1.25m Band

In the realm of Baofeng radios labeled 'Tri-Band,' a unique gem emerges in the form of the 220-250 MHz range, also known as the 1.25m Band within the Amateur Radio community. Nestled on the higher end of the VHF band, it extends toward the beginning of the UHF range. Renowned for its exceptional performance in woodland environments, surpassing UHF but falling slightly short of the lower VHF band, this band adds a layer of flexibility to radio communication.

However, the scarcity of radios operating in this frequency range bestows upon it an air of secrecy. The lack of attention to this portion of the spectrum can inadvertently enhance both the security against interception and exploitation. It's crucial to highlight that while this obscurity provides a degree of security, fundamental practices of Communication Security (COMSEC) should never be overlooked.

In Summary

The Baofeng radio stands as a beacon of versatility housed in an affordable package, empowering small groups with the skill to harness its potential effectively. This manual strives to unveil the radio's field usage with minimal infrastructure, crafting a robust communications plan, detailing communications techniques, offering insights into improvised antennas, and guiding the implementation of digital communications and encryption for maximum security. Emphatically, it's not a treatise on Ham radio but a practical guide on applying communications for guerrilla operations.

UV-SR'S BASIC CONTROLS AND MODES

Basic Controls

The UV-SR radio reveals a simplistic layout, with an omnipotent on/off volume knob, a keypad, a push-to-talk (PTT) button, and a set of unique-function buttons on the side. The entire radio, including embedded menus, is at the operator's command through these controls, offering a streamlined troubleshooting experience in the field, especially for seasoned operators. With each button having a maximum of two functions, corrective actions become straightforward.

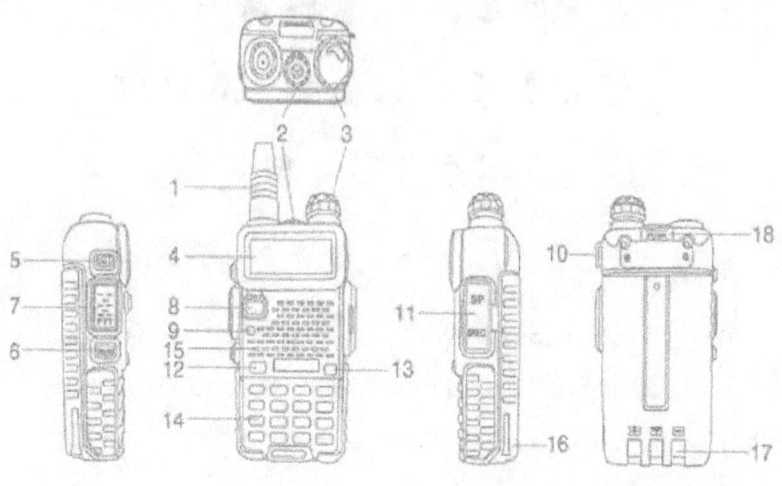

1. antenna
2. flashlight
3. knob (ON/OFF,volume)
4. LCD
5. SK-side key1/CALL(radio,alarm)
6. SK-side key2/MONI(flashlight,monitor)
7. PTT key(push-to-talk)
8. VFO/MR (frequency mode/channel mode)
9. LED indicator

10. strap buckle
11.accessory jack
12.A/B key(frequency display switches)
13.BAND key(band switches)
14.keypad
15.SP.&MIC.
16.battery pack
17.battery contacts
18.battery remove button

Frequency and Memory Modes

Upon unboxing, attention is drawn to the VFO/MR button below the screen. VFO, or Variable Frequency Oscillator, hearkens back to the era of crystal-controlled radios, where a crystal set the operating frequency. In contemporary terms, VFO translates to frequency mode, while MR signifies Memory or Channel mode, housing a list of pre-programmed frequencies. In Frequency mode, operators can directly input the desired frequency, while in Channel mode, they choose from the pre-programmed channels.

In Channel mode, small numbers on the right side denote the programmed channels. In Frequency mode, these numbers are absent. However, for radios using narrowband frequencies, two small numbers appear next to the large digits representing the frequency. It's crucial to grasp the distinction between 'channel' and 'frequency'—they aren't interchangeable. One refers to a specific operating frequency, the other to a frequency stored in memory or assigned to a specific band or radio service. Despite the complexity in terminology, understanding these nuances is pivotal.

Security in Memory Modes: A Cautionary Note

The data entered into these radios is a potential vulnerability. If captured, it becomes exploitable and can be used against the operator. Every adversary should be regarded as an intelligence collector, whether formally trained or not. Radios, like the UV-SR, are prized targets. In combat zones, adversaries with similar radios became subjects of interest. Recovered radios were not turned off immediately but handed to signals intelligence collectors.

The contents were meticulously analyzed to unveil the enemy's communication plan.

Under no circumstances should the memory of a radio intended for tactical or clandestine purposes be programmed. Compromise not only exposes you but also risks complacency within your group. Complacency doesn't just endanger lives; it erases hard-won ground. From a counterintelligence standpoint, the goal is to reveal as little as possible, creating an illusion of weakness and inferior capability.

Navigating the Baofeng Screen: A Practical Guide

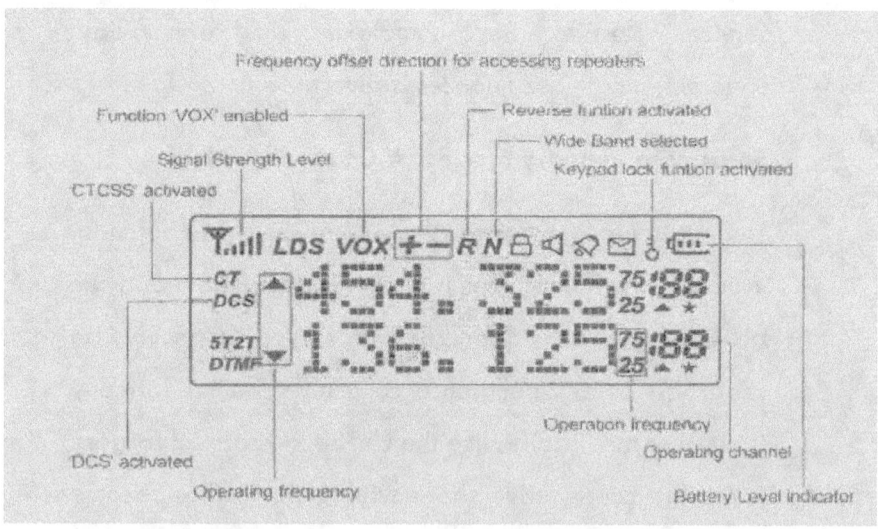

Understanding the Baofeng's LCD screen is essential for effective operation. Here's a comprehensive breakdown of visual cues and practical steps for entering or changing frequencies:

LCD Screen Overview:

The screen displays top and bottom frequencies, a battery life indicator, and a small signal strength meter in the top corner. Various symbols indicate enabled menu functions, such as VOX, power level, and repeater offsets. Arrows to the left of frequency numbers indicate the active frequency (top or bottom).

Channel Mode Display:

In Channel Mode, the radio can show alphanumeric letters, useful for naming channels. This feature is commonly used for labeling license-free frequencies, Amateur Radio or GMRS operations, or specific frequencies in a communications plan.

Entering or Changing a Frequency:

Keep the radio in Frequency mode by pressing the VFO/MR button. If voice prompts are enabled, it will indicate Channel Mode or Frequency Mode. Directly input the desired frequency using the keypad. Note that input requires all six digits for completion (e.g., 144.100). If an error occurs, press EXIT or wait a few seconds to revert to the previous frequency.

Channel Mode Entry:

In Channel Mode, input digits correspond to the channel number in the memory list. For example, enter 1 then 2 for channel 12, or 0 then 2 for channel 2.

Switching Frequencies or Channels:

Use the A/B button (number 12) to toggle between top and bottom frequencies or channels. The black arrow indicates the active frequency or channel.

Troubleshooting Tips:

If you encounter difficulties, give the radio a moment; it will revert to the previous operating frequency. Overthinking equipment can lead to frustration; often, issues stem from a lack of operator knowledge.

Remember, simplicity is key. The Baofeng's straightforward controls, when understood, empower operators to navigate frequencies and channels with ease.

Securing Communication

The Crucial Role of Keypad Lock in Radio Operations

In the intricate world of radio operations, where reliability can be a matter of life or death, ensuring seamless communication is paramount. A pivotal aspect often overlooked is the simple yet powerful act of locking the keypad. This practice, particularly crucial during nighttime patrols and other challenging scenarios, can prevent unintended frequency shifts, maintaining operational security and preventing potential friendly-fire incidents.

The Nighttime Patrol Incident: A Real-world Lesson

In a training scenario, a seasoned patrol leader, well-versed in conventional infantry tactics, found himself grappling with a new challenge — the Baofeng UV-SR radio. The patrol, split into assault and support-by-fire elements, moved with precision through the night. However, an unfamiliarity with the UV-SR's intricacies led to a critical error. The patrol leader failed to lock the control panel before movement, resulting in a frequency shift during the patrol. This seemingly minor issue cascaded into a loss of communication, and, in the absence of a locked keypad, a potential friendly-fire incident ensued.

This incident underscored a fundamental principle — in the realm of radio operations, every operator must possess an intimate understanding of their equipment. The UV-SR, though a reliable tool, demands a level of familiarity akin to one's proficiency with weapons. Without this understanding, the reliability of communication systems falters, jeopardizing the success of missions and the safety of operators.

Training Reality: Communication is Key

Training scenarios further accentuated the critical importance of communication, especially under low-visibility conditions. The ability to program and troubleshoot radios on the fly emerged as a vital skill. In operational environments devoid of external support, operators must exhibit comprehensive knowledge of their radios — a proficiency parallel to their mastery of weaponry.

The reality of training is stark — without effective communication, the likelihood of mission success diminishes significantly. It's not just about having radios; it's about knowing how to use them in dynamic and challenging situations. This reality extends beyond the training field, into operational theaters where real-world consequences hinge on the reliability of radio communications.

Establishing a Habit: Locking Down the Keypad

A simple yet effective habit emerged from these experiences — locking the keypad after each adjustment. Whether during movements or in moments of relative calm, this practice minimizes the risk of unintended frequency changes. It is a proactive measure, an operator's commitment to maintaining the integrity of their communication system.

Keypad Lock Procedure: Simple yet Impactful

On the Baofeng UV-SR, locking the keypad involves holding down the '#' key. This action prompts a voice prompt, if enabled, saying "LOCK," accompanied by a small key icon beside the battery indicator. With the keypad locked, all functions, except the push-to-talk (PTT) button, are disabled. To unlock, the operator repeats the process, holding down the '#' key until the voice prompt says "UNLOCK," and the icon disappears.

However, it's crucial to note that the lock function does not extend to volume control. A cautionary tale involves a Marine Communications NCO who, despite having a locked screen, lost communication due to an inadvertently turned-down volume knob. This incident serves as a reminder that comprehensive gear awareness is paramount.

Learn, Adapt, and Secure: Enhancing Operator Proficiency

Learning from experiences, both positive and challenging, contributes to operator proficiency. Locking the keypad is a straightforward yet impactful step in enhancing the reliability of radio communications. In the dynamic and unpredictable world of operations, securing your equipment is not just a precaution; it's a fundamental aspect of mission success and operator safety.

Unlocking the Baofeng's Versatility

Exploring Additional Functions

Beyond the basics, the Baofeng UV-SR radio harbors a spectrum of features that can enhance its utility for a range of scenarios. Unraveling these functionalities contributes to a more nuanced understanding of the radio's capabilities, enabling operators to leverage its potential to the fullest.

Top Side Key Functions (#5 in the Diagram):

- *Quick Press:* Switches to receive broadcast FM.

- *Hold Down:* Activates an alarm, transmitting the alarm tone on the set frequency. This feature is invaluable for quick alerts within a group or potentially disrupting enemy communication on shared frequencies. Caution is advised, as inexperienced users might trigger it accidentally.

Button Underneath the PTT (#6 in the Diagram):

- *Short Press:* Activates the LED light on top. Two quick presses turn it into a strobe. Operators often disable or camouflage this light, considering its potential to compromise stealth.

- *Long Press:* Activates the Monitor, opening the squelch. This is instrumental for receiving weak signals, allowing operators to hear transmissions even in challenging conditions.

Antenna Port and Microphone Jack:

- *Antenna Connection:* The Baofeng employs an SMA-Male antenna connection. While functional, it lacks ruggedness. Many users replace it with BNC adapters, offering durability and ease of use with various antennas.

- *Microphone Jack:* Located on the side opposite the PTT, it uses a Kenwood two-prong connector. This port serves dual purposes — facilitating microphone connectivity and acting as the data port for software programming and data bursts.

Quick Programming Menu Options (Selected Highlights):

Menu 0: Squelch

Filters weak signals to minimize static. Recommended setting is 1 for optimal reception.

Menu 1: Step

Defines the frequency spacing, crucial for scanning. Suggested setting is 6.25kHz.

Menu 2: TXP (Transmit Power):

Allows selection of transmit power. Consider low power for tactical and clandestine purposes.

Menu 4: VOX (Voice Activation):

Enables voice-activated transmission. Useful for digital operations but recommended to be off for other purposes.

Menu 7: TOR (Transmit / Dual Receive):

Essential for split mode operation, enabling reception on one frequency while transmitting on another.

Menu 8: BEEP

Control: Manages the keypad's beeping noise. Turning it off enhances stealth in operational settings.

Menu 9: TOT (Time Out Timer)

Function: Shuts off the radio after a designated time of transmitting. Setting it to 60 seconds prevents extended transmissions, avoiding accidental hot-miking.

Menu 18-13: DCS / CTCSS

Role: Governs sub-audible tones emitted and received by the radio. CTCSS aids in preventing communications jamming but can be easily detected. Setting them to OFF simplifies field operations.

Menu 14: Voice

Purpose: Manages the voice prompt during menu navigation or frequency input. Turning it OFF streamlines operation for tactical use.

Menu 23: BCL (Backlight)

Control: Manages screen backlighting. Turning it OFF is advised for tactical field use, minimizing visibility.

Menu 25: SFT-D (Standard Offset for Repeaters)

Usage: Sets the standard offset for repeater functions. For simplex operation (covered in this manual), set it to OFF if not using repeaters.

Menu 26: Offset

Function: Controls the offset frequency for repeater operation. Customizing this is crucial for groups establishing their own repeater systems.

Menu 29-31: LED Controls

Management: Governs backlight triggers during transmission or reception. Setting them to OFF aligns with tactical field use, minimizing unnecessary signals.

Menu 39: ROGER

Annoyance Control: Manages the roger beep, often considered an annoying feature. Setting it to OFF eliminates the post-transmission tone.

Menu 40: Reset

Critical Feature: Performs a factory reset, akin to military radio Z functions. Resets all settings, preventing potential exploitation. Use with caution, and reset ALL for a comprehensive reset.

In Summary

Understanding these menu functions facilitates efficient troubleshooting in the field. Consistent training is key to mastering the radio's simplicity. When issues arise, checking and adjusting these menu items according to the provided checklist can save considerable time and prevent headaches down the road. As with any critical task, practice and familiarity with the equipment ensure optimal performance. How have you utilized these menu functions in your operational context? Share your insights and experiences with Baofeng's inner workings.

COMMUNICATIONS PLANNING

Unveiling the Signals Operating Instructions (SOI)

Crafting effective communication within a group stems from meticulous planning encapsulated within the Signals Operating Instructions (SOI). This comprehensive chart structures a group's communications, fostering seamless coordination between teams operating in a designated area and their headquarters, the Tactical Operations Center (TOC).

Layers of the SOI

The SOI comprises diverse layers - frequency tables, team call signs, passwords, and an encryption grid known as the Search And Rescue Numerical Encryption Grid (SARNEG). Each layer bolsters communication security (COMSEC), safeguarding transmissions against potential interception.

Revisiting Sound Techniques

The global shift towards digital encryption and expediency, sidelining traditional methods, led to complacency and vulnerability. Rediscovering Cold War-era practices becomes imperative for those embracing the guerrilla role, promoting resilience against potentially superior adversaries.

Sensitive Nature of the SOI

The SOI, akin to the planned route, remains highly sensitive and is discarded after each mission. It's never reused, ensuring operational security. Units engaged in Unconventional Warfare in Vietnam adopted multiple frequency sets per mission, anticipating adversaries with advanced interception capabilities.

SOI Components and their Role

The SOI involves a PACE plan, delineating frequency sets, labeled with codewords, comprising transmitting and receiving frequencies. Operating through TOR/Menu #7 on the Baofeng configures these sets, adding a layer of protection against interception.

Elements of the SOI

The second phase of the SOI encompasses call signs, passwords, SARNEG, and Commo Windows, guiding communication procedures via radio and in person. Scrubbing the entire plan post-operation is paramount, preventing reuse and maintaining security.

Counter-Guerrilla Operations

Intelligence collectors prioritize recognizing patterns, considering the recovery of an SOI as a significant breakthrough. The SOI's strategic layout aims to challenge adversaries, emphasizing robust tradecraft to impede enemy efforts.

The provided Sample SOI Chart exemplifies the systematic layout, showcasing the critical elements necessary for effective communication planning, emphasizing the need for meticulousness and vigilance in tactical and clandestine operations.

Sample SOI Chart:

Signals Operating Instructions (SOI)

DURATION:

Primary [FREQ/ MODE]:

CODEWORD:

TX:

RX:

Alt. Freq [FREQ/ MODE]:

CODEWORD:

TX:

RX:

Contingency Freq:

Emergency Signal:

Callsigns:

TOC / Control:

Element 1:

Element 2:

Support/Recovery:

Challenge/ Password:

Running Password [DURESS]:

Number Combination:

Search and Rescue Numerical Encryption Grid [SARNEG]:

KINGFATHER

0 1 2 3 4 5 6 7 8 9

*** SARNEG first number rotates with last numeral of previous date***

COMMO WINDOW SCHEDULE:

1 2 3 4 5 6 7 8 9 10

• DAY:

• NIGHT:

Comprehensive PACE Planning

In mission planning, PACE, an acronym for Primary, Alternate, Contingency, and Emergency, forms the foundation by providing a four-tiered approach. This methodology extends to various mission facets, including route selection, equipment readiness, and communications. The robustness of these layers proves crucial not only for operational capabilities but also for ensuring communications security and the team's survival during personnel recovery phases.

Understanding the Role of Each Layer

- **PRIMARY:** The primary and most utilized method of communication.

- **ALTERNATE:** A backup option employed if the primary mode fails.

- **CONTINGENCY:** Activated to initiate personnel recovery plans.

- **EMERGENCY:** Non-electronic and aids in recognition for near and far linkup.

Both Primary and Alternate layers are largely interchangeable, featuring transmit and receive frequencies, along with the designated mode of operation. Clear markers like "voice" or specified digital protocols (e.g., MT-63, DMR, THOR) ensure operational clarity. Numerous uncontrollable factors, such as environmental changes or sophisticated electronic warfare by adversaries, may lead to communication failure. In such cases, swift transition to a new frequency or set is crucial. A code word uniquely identifies each frequency set, facilitating smooth transition upon code word exchange.

Example Implementation:

For instance, when transitioning to the alternate frequency set (assigned Codeword: ALICE):

- "Wolf, this is Hound, Over."

- "Hound, this is Wolf."

- "LAGOON."

- "LAGOON, Roger." *(Teams proceed to enter the new frequency set)*

Shift to Personnel Recovery (PR) Phase

Should both primary and alternate communication methods fail, the mission objective shifts towards personnel recovery. Historical trends indicate that communication failure often leads to mission failure or worse, loss of life. Contingency, reserved for PR, possesses a unique codeword for minimal verbal instructions over the network. An analogy can be drawn to the aviation term "MAYDAY," universally recognized and activated by a triple repetition on a specific frequency (e.g., 121.5 MHz), triggering immediate rescue efforts. This principle parallels the Running Password or Duress word within the SOI, functioning similarly for swift action in critical situations.

When strategizing missions in hostile territories like Iraq and Afghanistan, a contingency frequency was allocated, also known as the Medical Evacuation (MEDEVAC) frequency. This was a singular channel devoid of frequency hopping and encryption, ensuring unwavering reliability, especially crucial when lives hung in the balance. Consequently, the Contingency line within an SOI necessitates a single, steadfast frequency. Traffic on this frequency is rigorously limited since it's employed when the team is under duress or compromised, primarily utilized for linking up at the prearranged extraction site, activated with a pre-planned codeword. Precise and detailed planning becomes paramount in such scenarios.

In contrast, the Emergency signal is a non-electronic means used to communicate a team's location to the rescue unit. It incorporates Day and Night signals, commonly executed using tactical signaling tools like VS-17 flags during the day and colored chemical lights at night. These signals serve a dual purpose: marking the team's location and signifying they are not held captive. Upon spotting the signal, the responding team reciprocates, initiating Near and Far Recognition Signals, leading to the team's linkup, exchange of bona fides (challenge and password from the SOI), and their safe return to friendly lines.

Significance of Callsigns

Callsigns play a crucial role in network traffic identification, utilizing a combination of words, numbers, or letters to signify the source of network communication. In civilian contexts, they serve to distinguish individual stations, clubs, or events in radio communication. On the military front, callsigns convey an organization's identity and hierarchical authority. For instance, in a Light Infantry Company, APACHE designated Alpha Company, while numeric identifiers (APACHE 6, 5, 7, etc.) indicated specific roles within the hierarchy.

However, in unconventional warfare and guerrilla operations, concealing hierarchical structures becomes paramount. Efforts are made to obfuscate levels of authority and roles within the organization during communication, mandating frequent changes in callsigns for added security. I recommend employing a thematic approach for callsigns, ensuring relevance while obscuring actual positions of authority. Callsigns must distinctly represent three pivotal roles:

1. The Tactical Operations Center (TOC) acting under the Ground Force Commander's authority.

2. The Maneuver Elements (Teams) engaged in operations.

3. The dedicated Personnel Recovery (PR) and Support Element.

Security Measures

Passwords, Authentication, and Encryption

Understanding the strategic significance of each role, communicated during the operation's planning stage, is crucial, as is the concerted effort to veil the transmission's purpose from potential signals intelligence collectors.

Passwords and Authentication Techniques:

- Challenge and passwords are solely utilized in face-to-face encounters and are never articulated over radio channels. Their primary role involves identifying allies during field linkups. Even among familiar groups, these phrases serve as a vital signal for assessing the safety of a rendezvous. The exchange occurs when one person from each group meets, with one stating the challenge and the other responding with the password.

- This method, known as exchanging bonafides, involves using seemingly innocuous words or phrases devoid of meaning when isolated. For instance, a challenge and password such as "dog" and "horse" might be communicated as:

    - "Excuse me, did you see the lost dog poster?"

    - "No, I thought they were looking for a horse."

- In the event of the wrong password being given twice, any further attempts at linkup should be halted, and the compromised group must be deemed unsafe for contact.

Running Password and Distress Signaling:

- The Running Password, also known as the Duress word, signifies a team in distress. Its use prompts immediate action from the Tactical Operations Center (TOC) or Commander, shifting to the Contingency frequency to coordinate the team's recovery. Comparable to the common distress signal "MAYDAY" in aviation or maritime settings, it signifies an urgent situation but employs a different word in small group settings.

Search And Rescue Numerical Encryption Grid (SARNEG):

- The SARNEG serves as a robust method for authenticating transmissions over the air. Assigning each letter a corresponding number and rotating this sequence daily enables secure authentication during communication. It's versatile and changes daily, making it viable for multi-day missions.

- Utilizing the SARNEG involves initiating contact with the receiving station, using the callsign followed by the authentication code. For example:

  - "Hound, this is Wolf, over."

  - "Wolf, this is Hound."

  - "Authenticate KILO, ONE, over."

  - "I authenticate ZERO, INDIA, over."

- Beyond authentication, the SARNEG also facilitates encryption, particularly in disguising numerical information in messages. This method renders the information nearly indecipherable without prior knowledge of the current SARNEG sequence, significantly impeding signals intelligence efforts.

Communications Windows

During critical phases of movement, specifically insertion and extraction, teams operating on the ground face heightened vulnerability. When in motion, all radio equipment remains active and vigilant for potential enemy encounters. Once they secure their hide site, patrol base, or safe house, they send a safety report, switch off radios to conserve battery life, and limit communication to outside their hide site at a designated transmission site.

Purpose and Execution of Communications Windows:

- Communications Windows, commonly known as "comma windows," are prearranged time slots for teams to check in with the Tactical Operations Center (TOC) or other field teams. Usually, these windows occur every 12 or 24 hours, allocated in two-hour blocks, allowing flexibility for transmission within that period.

- In Afghanistan, two commo windows were a norm within a 24-hour cycle, strategically scheduled for daylight and night periods. Missing two consecutive windows would automatically trigger a personnel recovery plan, indicating potential trouble with the ground team.

- A personal anecdote highlights the significance: a team missed two windows due to inadvertent sleep after being exposed to cold, emphasizing the critical importance of adherence to these check-ins even for highly disciplined teams.

Communication Planning Insights:

- While these meticulous methods are tailored for high-threat environments, their application may seem excessive for standard communication needs. However, in tactical or clandestine settings, such detailed planning becomes imperative, irrespective of the enemy's sophistication. Failure to recognize this reality may lead to compromise and failure.

- The strategies outlined in planning SOI prevent the creation of predictable patterns. Regular alterations in practices confound predictive analysis, thwarting potential threats and ensuring survival.

Effective Communication Protocol

When it comes to radio communication, brevity is key. Keeping transmissions concise and clear enhances their effectiveness. Enunciation and precision are crucial to ensure the receiver comprehends the message accurately. There's a fundamental assumption while transmitting:

1. **Everything Monitored:** Assume all communications are being actively monitored.

2. **Everything Recorded:** Operate with the understanding that every word might be recorded.

In unconventional warfare settings, there's no room for routine transmissions. Breaking radio silence risks interception and detection, which is critical in tactical and clandestine operations.

Establishing Contact:

Initiating communication involves stating the callsign of the station being called, followed by one's own. For instance, "You, this is Me, Over."

Prowords for Efficient Communication

Prowords are concise terms with specific meanings aimed at minimizing unnecessary chatter. Examples and meanings include:

- **OVER:** Conveys the end of a statement, awaiting a response."*I'm done, your turn.*"

- **OUT:** Indicates completion of transmission."*I've finished transmitting.*"

- **ROGER:** Signals understanding or acknowledgment."*I understand.*"

- **COPY/HOW COPY:** Asks for confirmation or repetition of the message.""

- **BREAK:** Indicates moving to a new point or subject."*I am going to the next line/ I am breaking in*"

- **SAY AGAIN:** Requests a repetition of the previous message." *repeat what you said again*"

- **WILCO:** Confirms understanding and implies compliance." *I will comply*"

- **AUTHENTICATE:** Requests verification through a specific method."*I need the corresponding SARNEG combination*"

- **REPEAT:** Asks for the retransmission of a previous message." *Send the same firing solution (field artillery)*"

Considerations in Usage

While prowords ensure concise communication, their frequent use might reveal patterns to interceptors. Their significance lies in common practice, and for guerrilla tactics, alternatives might be necessary to evade detection while maintaining effective communication.

In the historical context of the first and second Chechen wars, Russian attempts at gathering significant intelligence from signals intercepted from Chechen leadership faced a formidable challenge. The Chechen leadership, predominantly comprising Soviet-Afghan war veterans, possessed prior knowledge of Russian capabilities and their failures. Leveraging this knowledge, they relied on local dialects and equipped villagers in targeted zones with commercial VHF radios. This flood of inconsequential communications inundated the airwaves, concealing crucial information amid a sea of trivialities.

The Chechen leaders' understanding of their own organizational needs and pitfalls led to their initial success. Employing local slang not only bolstered the morale of the people but also reinforced the struggle for survival, highlighting critical elements for successful guerrilla leadership.

NATO Standard Phonetic Alphabet:

Letter	Code Word	Pronunciation
A	Alpha	AL-fah
B	Bravo	BRAH-voh
C	Charlie	CHAR-lee
D	Delta	DEL-tah
E	Echo	ECK-oh
F	Foxtrot	FOKS-trot
G	Golf	Golf
H	Hotel	HOH-tell
I	India	IN-dee-ah
J	Juliett	Jew-lee-ETT
K	Kilo	KEY-loh
L	Lima	LEE-mah
M	Mike	MIKE
N	November	no-VEM-ber
O	Oscar	OSS-cah
P	Papa	PAH-pah
Q	Quebec	keh-BECK
R	Romeo	ROW-mee-oh
S	Sierra	see-AIR-ah
T	Tango	TANG-go
U	Uniform	YOU-nee-form
V	Victor	VIK-tah
W	Whiskey	WISS-key
X	X-ray	ECKS-ray
Y	Yankee	YANG-key
Z	Zulu	ZOO-loo

Number	Code Word	Pronunciation
0	Zero	ZEE-roh
1	One	WUN
2	Two	TOO
3	Three	TREE
4	Four	FOW-er
5	Five	FIFE
6	Six	SIX
7	Seven	SEV-en
8	Eight	AIT
9	Nine	NIN-er

Within the military sphere, the NATO standard for phonetics prevails, while in law enforcement and similar contexts, names are often used as substitutes for letters. Standardized phonetics significantly streamline transmission time by eliminating potential confusion on the airwaves. While using the NATO standard might reveal a degree of training and organization to signals intelligence teams, the benefits outweigh this potential risk. However, for guerrilla operations, unpredictability reigns supreme. Crafting an internal set of standardized phonetics within a group might be ideal. Yet, maintaining coordination with other guerrilla bands across a region could pose challenges unless a universally agreed-upon standard is established.

EFFECTIVE REPORTS

SALUTE and SALT

Returning to the notion that transmitted messages are logged, the transmission of mission-critical information becomes imperative. To convey intelligence concerning enemy forces, a simple and effective report format, known as SALUTE and its supplementary report SALT, is utilized.

SALUTE Format:

- S: Size (PAX and VICS - People and Vehicles)

- A: Activity (What are they doing?)

- L: Location of activity

- U: Uniform (How can we identify them?)

- T: Time of observation

- E: Equipment (Details on weapons, gear, and capability)

SALT Format:

- s: Size (PAX and VICS)

- A: Activity

- L: Location

- T: Time of observed activity

Every individual equipped with a radio should possess at least a basic familiarity with these report formats and their transmission methods. More eyes reporting mean better situational awareness. SALUTE reports offer a clear, concise format to relay crucial information, aiding analysts in constructing a more comprehensive understanding of an enemy's capabilities.

At first glance, SALUTE and SALT reports may seem similar, with significant overlap between the two. SALUTE reports serve as the initial and comprehensive report, establishing the baseline, while SALT reports supplement the SALUTE for any observed changes in the target.

Sample SALUTE Report Interaction:

- "Wolf, this is Hound, Over."

- "Hound, this is Wolf. Authenticate Kilo One, Over."

- "Wolf, this is Hound. Zero India, SALUTE Report, Over."

- "Roger, Standing by."

- Detailed observations given line by line with breaks indicating transitions.

- Use of SARNEG for location and specifying the time zone as LOCAL for time of observation.

Purpose of SALUTE and SALT Reports: Identification and Handoff The primary purpose of transmitting SALUTE and SALT reports is to equip intelligence analysts with the necessary information for accurately assessing observed enemy capabilities or determining whether they are indeed adversaries. For the ground team, the key consideration is determining the amount of information needed for positive identification (PIO) of a target when transferring this data to other groups.

Operational Use in Afghanistan: Surveillance and Reconnaissance

In Afghanistan, SALUTE and SALT reports were extensively utilized to transfer targets between teams operating within a specific area. During missions monitoring suspected Taliban villages (conducting patterns of life observations before potential attacks), SALUTE reports specifically detailed identifying characteristics of individuals of interest. This allowed seamless transfer of information between teams, maintaining constant vigilance on the target—a fundamental technique in surveillance and reconnaissance training.

The Importance of Accurate Reporting

Accurate reporting is crucial as guerrilla warfare relies heavily on mobile warfare concepts, seamlessly blending reconnaissance and combat roles while conserving resources. The SALUTE report serves a pivotal role in identifying both targets of interest and those to avoid. Attacking well-protected troops, like line Infantry units of an occupying force, especially in the early phases of guerrilla warfare, is generally ill-advised. Instead, targeting support troops offers a more lucrative and often successful strategy, impacting the morale of the larger force dependent on them. Hence, a guerrilla's training must heavily emphasize observation and discernment.

Medical Evacuation (MEDEVAC) Report

The effective handling of Medical Evacuation (MEDEVAC) requests is crucial in combat scenarios, ensuring the prompt and appropriate treatment of casualties. While the intricacies of care from the battlefield to the guerrilla hospital lie beyond the scope here, understanding MEDEVAC requests is vital.

For in-depth knowledge of care under fire and guerrilla hospital strategies, I suggest referring to COL Rocky Farr's "Death of the Golden Hour and the Return of the Guerrilla

Hospital" and "Where There Is No Doctor." Additionally, it's imperative for a successful guerrilla movement to provide basic medical training to combatants and recruit medical personnel for the welfare of individuals within their spheres of influence.

Notably, historical figures like Che Guevara, a trained doctor, showcased the power of administering medical aid to the impoverished to bolster favorability among the populace, a model embraced by various groups, including the US Special Forces and Civil Affairs.

In a guerrilla commander's planning, treating battlefield casualties should be a focal point for maintaining morale. Demonstrating care for fighters under duress significantly influences their belief in a cause. Therefore, a well-executed MEDEVAC request plays a vital role in ensuring appropriate patient care.

Clandestine Reports and the ANGUS Report

Clandestine communication involves sending specific reports.These reports, sent during designated communication windows, are crucial for relaying specific information. For instance, during extended missions in Afghanistan involving surveillance or targeting multiple subjects across a wide area, our team regularly sent reports to update our higher echelon on our operational status.

These report formats originated during the Cold War behind the Iron Curtain in Europe and in Southeast Asia, continuing to be utilized today. It's important to note that the names of these reports serve the purpose of identifying the type of report being sent to the Tactical Operations Center (TOC). They're not labeled with numbering systems but rather designated as AAA, BBB, and so forth, to ensure clear identification and prevent confusion.

ANGUS Report Structure:

1. **Date/Time Group (AAA):** Timestamp indicating the date and time when the report is transmitted.

2. **Team Status (BBB):** Code word or phrase used to identify the team's status.

3. **Location (CCC):** Details regarding the team's location, likely encrypted with a specific methodology such as SARNEG for security purposes.

4. **Deviations (DDD):** Notable changes or deviations from the planned or expected course of action reported by the team.

5. **Additional Information (EEE):** Any further pertinent remarks, observations, or details of interest.

This structured format ensures the transmission of critical information in the ANGUS Report, aiding in maintaining situational awareness and allowing for timely updates on team status, location, deviations, and additional noteworthy information during covert operations.

- The ANGUS Report serves to inform about the initial entry of a team into an area. During covert entries, like the ingress and egress phases of an operation, a team is at its most vulnerable, having limited defensible positions.

- For safety, all radios within the team are activated during ingress and egress. Once the team reaches its designated operational area, they transmit an ANGUS Report and power down all communication equipment to avoid accidental transmissions.

The ANGUS Report, named so for its function, acts as a critical update on a team's safe entry into an operational area. This protocol ensures that the team remains covert and reduces the risk of accidental exposure.

BORIS Report

The BORIS report represents a comprehensive intelligence report that goes beyond the immediate actions outlined in

SALUTE or SALT reports. While the SALUTE and SALT reports are designed for quick or nearly immediate action, the BORIS report takes a more expansive approach.

BORIS Report Structure:

1. **Intelligence Report (AAA):** Title or indication that this report contains intelligence-related information.

2. **Date/Time Group (BBB):** Timestamp indicating the date and time when the report is created.

3. **Date/Time of Observed Activity (CCC):** Timestamp specifying the date and time of the observed activity being reported.

4. **Location of Observed Activity (DDD):** Details about the specific location where the observed activity took place.

5. **Observed Activity (EEE):** Description or details regarding the activity observed, including any notable occurrences or events.

6. **Description of Personnel, Equipment, Vehicles, Weapons (FFF):** Detailed descriptions of individuals, equipment, vehicles, and weapons observed during the activity.

7. **Team Assessment (GGG):** An evaluation or analysis conducted by the team regarding the observed activity and its implications.

This structured format ensures that essential intelligence-related details are conveyed consistently in the BORIS Report. It allows for a comprehensive understanding of observed activities and their potential impact on ongoing operations or strategic planning.

Key Characteristics of the BORIS Report:

- **Long-Term Observations:** It involves in-depth intelligence gathered from prolonged observation and analysis by the team on the ground.

- **Broader Scope:** Unlike the SALUTE report, which focuses on immediate actions, or the SALT report, which addresses specific details, the BORIS report covers a larger area. It's designed for long-term assessments and coordination of actions among multiple groups.

The BORIS report serves as a valuable tool for sharing comprehensive intelligence, facilitating coordination among different groups operating over a broader geographical region. This detailed assessment aids in strategic planning and decision-making, considering a larger operational landscape.

CYRIL Report

The CYRIL Report serves as a concise team status update and is transmitted at the onset of each communication window. It holds significance both for the team operating in the field and for the Tactical Operations Center (TOC) to maintain their situational awareness.

Key Attributes of the CYRIL Report:

- **Team Status Update:** This report provides a snapshot of the team's current status, possibly encompassing elements such as personnel count, operational conditions, equipment status, or any relevant developments.

- **Regular Transmission:** Sent at the commencement of each communication window, ensuring that both the team in the field and the TOC are regularly updated on the current situation.

By transmitting the CYRIL Report at the beginning of each communication window, it ensures that both the field team and the TOC are on the same page regarding the team's status and any recent developments. This synchronization aids in better decision-making and situational management.

CYRIL Report Structure:

1. **Situation Report (AAA):** A brief overview or summary of the current situation or status.

2. **Date/Time Group (BBB):** Timestamp indicating the specific date and time when the report is generated.

3. **Current Location (CCC):** Details about the team's current or updated location.

4. **Medical Status (DDD):** Information regarding the medical condition or status of team members.

5. **Equipment Status (EEE):** Report on the condition and functionality of equipment used by the team.

6. **Supply Status (FFF):** Status update on essential supplies such as batteries, ammunition, water, and food.

7. **Team Activity since last Commo Window (GGG):** Summary of team activities or operations since the last communication window.

8. **Team Activity until next Commo Window (HHH):** Planned or anticipated activities until the next communication window.

9. **Remarks:** Additional notes, observations, or remarks relevant to the current situation.

CRACK REPORT

CRACK is traditionally employed to assess battlefield damage to infrastructure. However, it serves a dual purpose of determining whether a facility can support the operational needs of a team on the ground. Forward surveillance teams often use CRACK to evaluate conditions of crucial points such as bridges and roads, crucial for follow-on forces during an offensive.

In specific operational scenarios like in Afghanistan, CRACK was utilized not only to locate potential staging points or safe houses but also to request necessary assets to enhance their capabilities to suit operational requirements. These assessments covered various aspects, including natural defenses, sanitation considerations, topography for signals intelligence collection, etc.

For potential guerrilla groups, CRACK can similarly serve in identifying locations that align with the group's needs for shelter, operational viability, and strategic advantages.

CRACK Report Structure:

CRACK: Battlefield Damage Assessment

1. **Date/Time Group (AAA):** Timestamp indicating the date and time when the report is created.

2. **Type of Target (BBB):** Categorization or classification specifying the type of target or infrastructure assessed.

3. **Description of Target (Physical and Functional Damage) (CCC):** Detailed description of the observed physical and functional damage to the assessed target or infrastructure.

4. **BDA Analysis/Resource Requirements (DDD):** Analysis of the Battlefield Damage Assessment (BDA) and identification of resource requirements necessary for repair, improvement, or utilization of the facility for operational purposes.

This structured format allows for detailed assessment and analysis of infrastructure damage or suitability for operational use, facilitating decision-making regarding resource allocation and strategic utilization.

UNDER REPORT

The UNDER report, also known as a Cache Report, serves as a means to report the location and details of caches. Guerrilla fighters, often limited in what they can carry over extended periods, rely significantly on hidden caches for supplies, equipment, or resources. These caches allow for strategic resupply without the need for extensive supply lines, reducing the visibility of the guerrilla fighter.

Historically, various guerrilla movements and clandestine activities have relied on such caches for operational sustainability, whether during conflicts like the Spanish Civil War, Nazi-occupied France, or activities conducted by intelligence agencies worldwide.

UNDER Report Structure:

UNDER: Cache Report

1. **Date/Time Group (AAA):** Timestamp indicating the date and time when the report is created.

2. **Type (BBB):** Classification specifying the type or category of cache.

3. **Contents (CCC):** Detailed list or description of the contents stored within the cache.

4. **Location (DDD):** Specific details or coordinates indicating the location where the cache is concealed or hidden.

5. **Depth (EEE):** Information regarding the depth or depth-related details concerning the placement of the cache.

6. **Additional Info/Reference Points (FFF):** Any supplementary information or reference points aiding in locating or identifying the cache.

The UNDER report provides a structured means of reporting the existence and details of these caches, aiding in their retrieval or utilization when necessary.

Absolutely, the provided report formats serve as a comprehensive guideline illustrating the structure and key components typically involved in clandestine communication reports. These formats offer insights into the methodologies utilized within the realm of covert activities, drawing from the practices followed by Western Conventional and Special Operations Forces.

However, it's crucial to recognize that these formats aren't rigid or absolute standards but rather a distilled product of institutional knowledge and experience. They provide a logical framework for conveying vital information effectively. How your group adapts and utilizes these formats should align with your specific operational needs and preferences.

The ultimate goal remains the same: conveying clandestine information in a concise and effective manner. It's imperative to prioritize the success and safety of your movement, ensuring that the information exchange process remains secure, efficient, and aligned with the operational objectives.

Flexibility in utilizing these formats while maintaining their core essence is key, allowing for adaptability to varying circumstances and operational requirements.

The Antenna: The Unsung Hero of Radio Communication

When newcomers venture into the world of radio operation, their initial queries often revolve around which make or model of radio to procure. Surprisingly, the fundamental component that often goes unnoticed or underestimated is the antenna. Ironically, it holds the utmost importance in the efficacy of radio communication.

Understanding antennas can be daunting; it's a realm that commonly falls into two extremes—either considered overly complex and disregarded or broken down into intricate engineering formulas that seem overwhelming. This section aims to bridge this gap.

For a radio operator, whether part of a guerrilla force or a small unit, grasping the antenna's role is as vital as understanding their rifle or any other essential equipment. The operator's safety and success hinge on this understanding.

This segment will delve into antennas, beginning with an exploration of the individual perspective—a closer look at the antenna attached to the radio. It will then broaden the discussion, encompassing both omni-directional and directionally transmitting antennas, offering insights into their relevance and usage for small unit requirements.

Crucially, this segment strives to present these concepts in a manner accessible to anyone navigating an improvised environment. It aims to highlight the practical significance of antennas without delving excessively into technical jargon.

Understanding the antenna's significance is pivotal for effective communication in the field, enabling operators to optimize their radio's capabilities and contribute to mission success.

Optimizing Baofeng Radio for Field Use

In various roles—Sustainment, Tactical, and Clandestine—the Baofeng radio serves adequately when certain considerations are accounted for. However, its stock configuration limits its utility, mainly due to the generally subpar factory antenna. Most users opt for aftermarket antennas to enhance performance.

The Baofeng utilizes a small, somewhat fragile SMA female adapter, prone to wear and breakage, particularly when used with larger aftermarket antennas. This issue isn't exclusive to Baofeng but applies to any radio employing SMA connections. My personal solution involves converting these connections to the more durable BNC connectors, offering better resilience in field conditions. The twist on/twist off nature of BNC connectors simplifies and secures quick connections, especially when attaching the antennas discussed later in this chapter.

For tactical use within a team, I prefer flexible antennas that deliver reliable transmission and reception performance. When incorporating a radio into your fighting load, positioning it on your non-primary side is crucial to avoid interference with the rifle stock. Personally, I weave the antenna into my shoulder strap's webbing rather than using an antenna relocation kit. The latter can be fragile and might disrupt antenna performance due to interference with the body's electrical field, a point further elaborated on in this chapter.

While aftermarket antennas offer some improvement, their utility is limited. Especially for roles beyond the tactical and inter-team levels, constructing field expedient antennas becomes essential. The subsequent part of this chapter will demystify the underlying theory, providing simple instructions for crafting common antennas necessary to maximize Baofeng's (or any other radio's) capability in challenging environments.

This section aims to equip the reader with the fundamental knowledge to construct various antennas, crucial for optimizing radio performance in austere settings.

Understanding Radio Wave Propagation

When exploring the range of handheld radios like the Baofeng, it's crucial to grasp some fundamental radio theory rather than asking solely about their range. Essentially, the range depends on several factors:

1. **Operating Environment:** The surroundings in which you're using the radio.

2. **Obstacles:** Anything blocking the signal path between you and the receiver.

3. **Antenna Efficiency:** The quality and efficiency of your radio's antenna.

A handheld radio primarily operates within the VHF and UHF ranges, generally operating on a "Line Of Sight" (LOS) basis. This implies that, theoretically, if one radio can visually "see" another on the radio horizon, communication between them is possible, irrespective of other factors.

Typically, stock antennas on handheld radios serve as a compromise, ensuring the radio doesn't overheat during transmission and providing enough signal quality to cover short distances, often just a few miles. Many users replace these stock antennas with aftermarket models for slight improvements, especially at the inter-team tactical level.

However, purpose-built antennas tailored to specific tasks offer the best performance. This chapter focuses on basic instructions for improvising antennas with available resources in any environment. This approach significantly enhances the capabilities of equipment like the Baofeng, contributing to the Communication Security (COMSEC) needs of Tactical and Clandestine roles.

Understanding improvised antenna construction involves delving into basic physics in simple terms. Visualizing radio waves and their propagation is akin to envisioning light. Imagine being in a dark room where a lightbulb suddenly turns on. This lightbulb represents your radio and antenna when transmitting, and the illuminated area symbolizes where your radio waves propagate or can be received. If you can imagine this scenario, you've taken a significant step toward comprehending radio theory.

Environmental Effects on Radio Propagation

The environment significantly impacts how far a radio wave can travel. For any communication purpose—be it Sustainment, Tactical, or Clandestine—understanding the operating environment is crucial to comprehend the potential limitations and strengths.

The general rule is that VHF signals (30-300mHz) perform better in rural, hilly environments, while UHF (300-3000mHz) works better in urban areas. This occurs because VHF ground wave signals physically bend with the terrain, much like how light behaves. Small terrain features like hills cast a smaller shadow, refracting signals in a way akin to light scattering off a mirror. Conversely, UHF signals are blocked by minor terrain features but can penetrate buildings.

Considering COMSEC (Communication Security), a team might use UHF for close inter-team communications, limiting potential interception over short distances. For longer-range communications with the TOC or a Guerrilla Commander, VHF, coupled with a purpose-built antenna, makes sense in rural settings. Conversely, in urban environments, the approach would be the opposite.

Further environmental impacts on radio waves include:

1. **Vegetation's Effect:** Higher frequencies degrade in dense vegetation like jungles, significantly shortening the radio signal coverage range. VHF typically performs better in woodland environments.

2. **Desert and Arctic Reflection:** Desert areas with minimal vegetation and highly reflective soil cover signals over longer distances. Similarly, snow and ice-covered Arctic environments exhibit comparable radio propagation performance.

3. **Water Bodies as Reflectors:** Bodies of water serve as excellent reflective surfaces for radio signals, enhancing coverage.

4. **Soil Reflectivity:** Soil rich in minerals like silicates or granite offers better radio wave propagation. Adding water underneath the antenna can enhance soil reflectivity; historically, even urination beneath antennas improved signal quality.

Understanding these environmental impacts is crucial for optimizing radio communications in different terrains and ensuring effective and secure transmissions.

Obstacles and Radio Transmission

The presence of physical obstacles between the sender and the intended recipient significantly influences radio communication. Devices like the Baofeng, operating in the VHF/UHF bands, rely on Line of Sight (LOS) communication. Essentially, if one radio can transmit RF (radiated energy) and 'see' the other, communication can occur. However, obstacles disrupt this process.

Large hills, ridgelines, and urban structures pose considerable challenges in maintaining communication. To overcome this, increasing the physical Line of Sight becomes essential. For instance, antennas often sit atop high ridges in mountainous terrains, while tall buildings in urban areas frequently host antennas. The higher the physical elevation, the greater the Line of Sight, allowing radio waves to cover more distance—referred to as groundwave.

In simple terms, when both the sender and the intended recipient are higher above the physical horizon line, their lines of sight extend further, enabling radio waves to cover more distance irrespective of the power output.

Obstacles in the environment also scatter signals through refraction, similar to how light reflects off a mirror. This aspect is vital in clandestine communication methods using directional antennas, as it plays a crucial role in managing and directing signals.

Understanding Antenna Efficiency

When discussing antennas, people often focus on the visible part sticking out of the radio, neglecting the equally crucial internal metal body—the ground. All antennas are fundamentally dipole antennas, comprising two equal and opposite halves: positive (the physical antenna) and negative (the built-in ground or reflector in the radio's metal body).

For a handheld radio, the physical antenna acts as the positive half, and the radio's built-in ground serves as the negative half or reflector. The closer these two halves match each other, the more efficient the antenna. An optimum antenna length, known as resonant, provides 2.15 dB (decibels) of gain. dB of gain measures signal strength, similar to candlepower or lumens for light.

Antennas exhibit gain or loss, where the further they are from a resonant dipole, the more loss they experience. A resonant dipole provides 2.15 dB of gain, effectively doubling the signal strength without requiring more power. Gain operates exponentially: every additional 3 dB of gain doubles the power again.

Calculating the physical length of the antenna involves a simple formula for a half-wavelength dipole:

FULL WAVELENGTH: 936 / Frequency in MHz = Physical Length in Feet

However, antennas don't operate in full wavelengths; they are half-wavelength dipoles. A half-wavelength is considered ideal for efficiency. The formula for this is 468 divided by the frequency in MHz:

HALF WAVELENGTH: 468 / Frequency in MHz = Antenna Length in Feet

For example:

- **VHF: 468 / 151.820 = 3.08 feet**

- **UHF: 468 / 462.550 = 1.01 feet**

This calculation provides the antenna length needed for optimal performance at the specified frequency.

To measure and cut the elements for each half of the dipole antenna, we divide 468 by two, resulting in 234. This gives us the formula for each antenna element:

QUARTER WAVELENGTH: 234 / Frequency in MHz = Length in Feet

For instance:

- **VHF example: 234 / 151.820 = 1.54 feet**

- **UHF example: 234 / 462.550 = 0.506 feet (6 inches)**

As a practice, it's advisable to confirm accuracy by calculating the quarter wavelengths for both elements and combining them, then calculating the half wavelength to ensure consistency. This verification helps identify any math errors before cutting materials.

It's notable that antennas become notably smaller as the frequency increases. Antenna length inversely relates to frequency—lower frequencies require longer antennas, while higher frequencies demand shorter ones. Despite UHF antennas potentially not performing as well in rural settings, their smaller size offers a lower signature and greater portability, making them advantageous in certain scenarios.

Types of Wire for Antennas

Utilizing various wire types for antennas is advantageous, simplifying wire acquisition in diverse settings, whether rural or urban. However, certain factors need consideration:

1. **Bare Wire Vulnerability:** Bare wire may ground itself when in contact with vegetation.

2. **Solid Wire Drawbacks:** Solid wire often breaks or dislodges from crimps.

Personally, I favor using electric fence wire for antennas due to its affordability, accessibility, and disposability. Yet, its bare nature presents a drawback—upon contact with a grounded object, it itself becomes grounded, potentially causing radio damage. Additionally, its solid structure makes it susceptible to breaking or coming loose from crimps.

Other wire options to explore include THHN (stranded electrical wire), speaker wire, and lamp cord. Despite being more expensive, they consist of small bundled strands coated with rubber or plastic insulation, enhancing their durability in field conditions.

In practical scenarios, makeshift antennas have been fashioned from Claymore wire (aluminum stranded lamp cord) and WD-1 commo wire (slash wire for field phones). However, lighter wires are more prone to breaking. Optimal tensile strength versus weight lies within the 16-14 gauge wire range.

Insulators for Antenna Ends

Insulators are affixed to the ends of antenna elements (wire segments), serving as endpoints and attachment points for securing lines or ropes for hoisting. Notably, 550 cord doesn't function as an insulator when wet, becoming electrically conductive and unfit for use as an endpoint insulator.

Field expedient insulators are easily sourced in various environments, ranging from plastic or glass pieces from bottles to wood in desert areas (less suitable in humid or jungle regions), or even plastic spoons with holes for wires and securing lines. Personal favorites include electric fence insulators made of plastic or ceramic, emphasizing that effectiveness doesn't necessitate high cost or aesthetics— practicality reigns in field expedient solutions.

Attaching the insulator involves bending a loop (referred to as a bite, akin to mountaineering) and securing the insulator at the end. This loop completion, whether bent or crimped, directs RF energy to the loop's apex rather than the wire's end.

Antenna Alignment and Designs

The orientation of antennas determines their polarization. There are two primary orientations:

1. **Vertical Polarization**: Antennas oriented up and down.

2. **Horizontal Polarization**: Antennas aligned parallel to the Earth's surface.

Most handheld radios and line-of-sight (LOS) radios are vertically polarized, indicating that their antennas are oriented vertically. A significant difference of 12dB in signal strength exists between different polarizations. When transitioning from vertical to horizontal polarization, the reception of signals may or may not occur with a vertically polarized antenna. This polarization aspect holds crucial significance for Communications Security (COMSEC). If search assets use vertically polarized antennas while your signal is horizontally polarized, they might entirely miss detecting your transmission.

However, LOS radios often employ vertical polarization due to their Frequency Modulation (FM) operation, as proximity to the earth's ground can scatter signals and cause signal loss.

While this isn't universally applicable, recognizing this tendency can be advantageous in certain situations.

Improvised Antenna Designs

Field expedient antenna designs cater to sustaining communications or expanding the communication range of a patrol, enabling both simple construction and high performance. These designs categorically fall into two types: omni-directional and directional antennas.

- **Omni-directional Antennas**: Radiate RF energy equally in all directions, akin to a lightbulb illuminating a room uniformly.

- **Directional Antennas**: Emit RF energy primarily in a specific direction, similar to a flashlight beaming light in one direction. This characteristic makes them proficient for COMSEC, as interception becomes considerably challenging unless a receiver aligns along the radiated path. Additionally, directional antennas possess high gain values, efficiently extending the operating range in the directed path. Analogous to how a flashlight illuminates a specific area strongly and extensively compared to an omni-directional light source, directional antennas serve the same purpose in the RF spectrum.

The Jungle Antenna: An Efficient Field Antenna

The Jungle Antenna, known by various names like the 292 Antenna, groundplane antenna, and military designation OE-254, is a straightforward and vertically polarized field expedient antenna. This antenna provides a 6dB gain in all directions. Its origins trace back to pre-World War II, during US forces' jungle combat training in anticipation of potential Pacific warfare. As radio technology emerged on the battlefield since World War I, advancements necessitated the best possible implementation. While training in Panama, the US Army Signal Corps found that modifying a vertically polarized dipole by adding two extra legs to the negative side, forming a pyramid shape, performed well in dense vegetation.

Construction of the Jungle Antenna, and indeed any antenna, can be made easier with a Split Post BNC Adapter, also known as a **Cobra Head**. These adapters feature a positive (hot) and negative (cold) terminal, indicated by red and black screw-down caps, respectively. They facilitate easy attachment of coaxial cable and its connection to the radio. To create the Jungle Antenna, cut four quarter-wave lengths of wire based on the operating frequency. One wire attaches to the Hot end, while the remaining three form a pyramid on the Cold side. Often, three sticks are tied together to create a triangular base for the cold end, establishing the groundplane.

Subsequently, attach the coaxial cable. Most transceivers use 50-ohm coax, and for field expediency, any 50-ohm coax cable will suffice. Although some loss might occur, it doesn't significantly affect operations unless exceptionally long cable runs are used. Typically, lengths under 18 feet work well. Common cable types include RG-58 and RG-8X, both equipped with BNC connectors, ensuring quick, secure, and straightforward attachment to the radio.

The primary purpose of the Jungle Antenna is to maximize signal range from a fixed location. For sustenance communication purposes, it's highly effective in establishing communication during the recovery phase of a natural disaster, extending the RF range to approximately 6 miles. In tactical settings, this radio would be stationed at the TOC for transmission, receiving messages from a dedicated transmission site during a patrol.

Directional Antennas for Guerrilla Communications

Directional antennas, discussed earlier in this chapter, transmit radio signals along a specific aimed direction, known as an azimuth. To effectively use these antennas, the guerrilla force operating in an area must first mark crucial locations on their map, such as the TOC (Tactical Operations Center) or Safe

House, and then identify planned transmission sites within their observation area. They determine azimuths from the TOC to these planned locations and calculate the back azimuths (by adding or subtracting 180 or 360 degrees in a circle) from the planned transmission sites back to the TOC. Using these directional antennas, they securely send and receive radio traffic, employing the report formats and encryption techniques discussed in the previous chapters. This tactical approach minimizes interception risks and the ability to determine the transmission direction (Direction Finding - DF), thereby safeguarding the team's operations.

The Sloping Vee: A Simple and Effective Directional Antenna

Among directional antennas, the Sloping Vee stands out as the most straightforward design. It essentially consists of a Dipole configuration with its two legs set at a 45-degree angle from each other, aligning the opening of this angle toward the intended transmission direction. Its simplicity facilitates quick construction, hoisting, and easy adjustments as needed.

Adding a carbon resistor of 500 ohms or more can potentially enhance the forward pull of RF energy. However, the sloping configuration of the dipole's legs inherently creates a radiating pattern moving forward in the intended direction, obviating the necessity for additional resistors for basic functionality.

The Yagi Antenna: Its Origin and Functionality

The Yagi antenna, perhaps one of the most recognizable directional antennas, holds historical significance and practicality in its design. Named after one of the physicists involved in its creation for Imperial Japan before World War II, this antenna was initially designed for the Imperial Japanese Navy's exploration of primitive RADAR. It operated by directing RF energy along a path and listening for reflections, which they utilized for communications security, making interception outside the specific signal path extremely challenging. Consequently, this method became integral for communication between occupied islands in the lead-up to the Pacific theater of World War II.

Post-war, the Yagi antenna gained widespread use in the US with the rise of over-the-air TV. Commonly affixed to household roofs with a rotor-controlled dial for optimal TV station reception, its operational concept parallels the Yagi's functionality.

Yagi Antenna Structure and Elements

At its core, a Yagi antenna is a simple Dipole, labeled here as the Driven Element (DE), augmented with additional components. It consists of the Driven Element (DE), a slightly longer Reflector (R), and a Director (D), which functions similarly to an aiming sight's front post. This specific Yagi configuration, known as a three-element Yagi, offers a gain of 7.5db.

Elements Spacing and Calculation

Spacing between the elements is crucially important, magnetically related, and labeled in fractions of a wavelength on the boom—the insulated central piece running through the antenna structure. These fractions, when multiplied by 936, yield the constant length in feet required for proper element spacing.

For instance:

- **Distance between R and DE: 0.18 wavelength (936 × 0.18 = 168.48)**

- **Distance between DE and D: 0.15 wavelength (936 × 0.15 = 140.4)**

Construction Process

Constructing a Yagi involves careful calculations and material selection. Beginning with determining the total boom length needed, one can use various insulating materials such as PVC, fiberglass fencing rods, or dry wood for this purpose. After calculating the total length required, marking off the placement of the Reflector, followed by the Driven Element and Director, guides the process from back to front in assembling the Yagi antenna.

Once the boom length is determined, the process continues with wire cutting. Copper welding rods are also excellent for crafting Yagi antennas.

For the Reflector (R), Driven Element (DE), and Director (D), specific calculations determine their lengths concerning the operating frequency:

- **R: 510 / Frequency = Length in Feet**

- **DE: 468 / Frequency = Length in Feet**

- **D: 425 / Frequency = Length in Feet**

The Driven Element is essentially a Dipole. To facilitate rapid assembly, the Cobra Head serves as the feed point, and electrical tape secures the Yagi elements in place. This straightforward method works effectively.

During an RTO Course in Wyoming, a Yagi antenna, readied for use with a camera tripod, was successfully constructed. It facilitated a 35-mile contact via voice and digital data burst to another Yagi positioned on a distant ridgeline within the class.

The Yagi's capabilities can be expanded by adding more directors to the front of the boom, enhancing gain and tightening the transmitted signal's beam width. However, a 5-element Yagi, offering 9.5db of gain, amplifies both transmission and reception signals significantly. This intensified reception might overwhelm the radio receiver akin to someone shouting loudly in your ear compared to a regular conversation. Hence, for practicality and effectiveness, sticking to the basic 3-element Yagi proves highly efficient.

Understanding SWR and Its Importance

Field expedient antennas, while not always aesthetically pleasing, are crafted from readily available materials, often performing as well as or even surpassing commercially manufactured counterparts. However, ensuring their efficiency demands attention to the Standing Wave Ratio (SWR), a critical measure of an antenna's effectiveness.

SWR quantifies the proportion of energy transmitted forward compared to what reflects back to the radio. Ideally, a perfect SWR is represented as 1:1. Antennas with an SWR below 2:1 are deemed usable. Conversely, an SWR above 2:1 poses risks, potentially damaging the transmitter due to heat buildup and risking harm to the final transistor. A dedicated SWR meter is an invaluable tool and should be included in a Guerrilla support group's toolkit for precise measurements.

Given that Baofeng radios lack built-in SWR protection, meticulous attention to element measurements during the construction of field expedient antennas is vital. Greater precision in measurements results in closer resonance, approaching the coveted 1:1 SWR. Achieving this near-perfect SWR significantly enhances transmission efficiency, a factor that could prove crucial in critical situations, possibly determining life or death scenarios.

COMMUNICATION STRATEGIES

Communication serves three key roles - sustainment, tactical, and clandestine, each demanding specific approaches for successful operations. It falls upon the Guerrilla Commander and communication personnel to discern these roles at different times. Failing to acknowledge this principle can result in a critical failure regarding communications security (COMSEC).

Personal Considerations

The use of communication devices should be restricted to key leaders and specifically trained communication specialists. COMSEC is as crucial as understanding the mechanics of communication. A tool like the Baofeng radio can be life-saving, but mishandling it can lead to fatal consequences. Radios must be handled with the same discipline as handling a rifle, ensuring minimal noise and light during patrols. In modern conflict zones like Afghanistan, Iraq, and Ukraine, meticulous electronic spectrum surveillance has been vital. The best strategy for a Guerrilla against technologically advanced foes is to restrict transmissions based on discussed principles.

Leaders and dedicated communication experts equipped with radios have specific reasons for transmitting messages. Unnecessary or incorrect transmissions must be minimized. Crafting a script before transmission helps reduce errors like excessive filler words or mistakes during message transfer. By adhering to these fundamental practices, messages remain concise and precise.

Sustaining Communications

Sustaining communications, aims to establish regular communication channels in scenarios where such communication might not exist. It comes into play post-natural disasters or when oppressive governments shut down communication networks.

Sustaining Communications in Guerrilla Warfare

A Guerrilla Commander's core concern isn't merely combat; it's about winning the people's support. Guerrilla groups aim for social reform amidst oppression or injustice. Historical examples demonstrate the pivotal role that sustaining communications plays in guerrilla warfare.

Adapting the Baofeng radio for this role is straightforward— it becomes a vital conduit for information in otherwise isolated areas. Employing a standard SOI (Signal Operating Instructions) and an omni-directional antenna enables rapid establishment of local area communications. Evaluating the physical reception range involves distancing radios and marking their locations on a map. Typically, a Jungle Antenna paired with a Baofeng offers a realistic range of 6-10 miles, subject to terrain and frequency variables. Although repeaters can extend the reach, they aren't dependable for tactical purposes. Taliban experience showcases how analog repeaters were easily compromised, leading them to prefer HF radio.

Strategically dispersing spare Baofeng radios and SOIs among households creates an independent communication infrastructure, earning favor among the local populace and aiding in medical aid, targeted force coordination, and morale improvement.

Tactical Communications

Using the Baofeng tactically requires strict adherence to COMSEC. Key leaders are the only ones entrusted with radios to minimize unnecessary chatter.

Tactical radio use serves the purpose of coordinating maneuvers and relaying battlefield intel. It aids in coordinating movement and preventing friendly fire, especially in low-visibility scenarios and dense environments.

Principles for Baofeng use in tactical settings:

1. Keep transmissions brief, employing code words when possible.

2. Employ separate receive and transmit frequencies.

3. Reduce power and use shorter antennas to limit transmission range, minimizing interception and direction finding risks.

4. Avoid transmitting atop terrain features; remain in valleys or below ridge crests to mask signals and complicate direction finding.

5. Use the least optimal frequency range for the terrain (UHF for rural, VHF for urban) to restrict the reach of radio traffic.

Facilitating a Tactical Operations Center (TOC)

The essence of a Tactical Operations Center (TOC) lies in directing and orchestrating ground-level tactical elements. Unlike a static position, a TOC must be adaptable and flexible. During extensive operations in Afghanistan, the Ground Force Commander adeptly ran a mobile TOC from a vehicle, coordinating activities spread across vast distances despite geographical challenges. Despite multiple terrains and obstacles, his group functioned as a mobile TOC while reporting to higher echelons. A Guerrilla TOC operates similarly, requiring:

- Radio for team communication

- Omni-directional antenna for extended radio range

- Tools for recording and displaying reports

- Area maps

- (Optional) A coffee maker for operational continuity

While employing an omni-directional antenna maximizes RF signal range and reception capability in various directions, unlike the Afghanistan scenario, Guerrillas must be vigilant of interception threats and their high-priority target status. To counter this, rigorous Signal Operating Instructions (SOI) planning and adherence to COMSEC protocols are paramount.

Clandestine Communications

Clandestine communications often instigate tactical actions, either based on intelligence or in response to a compromise. These transmissions relay information gathered using report formats ,termed "Products" in intelligence contexts.

Such communications, usually sent over extended distances during specific communication windows, necessitate directional antennas. In mission planning, patrols, regardless of size, identify transmission sites and note the azimuth to the TOC for transmitting encoded messages .As a guideline, no radio transmissions originate from the Hide Site, Patrol Base, or Safe House. Transmission sites, at least 1000m away from these locations, protect team operational security. This standard practice among Cold War-era Unconventional Warfare (UW) units aims to prevent additional casualties if transmissions are intercepted, located using Direction Finding (DF) methods, and targeted with artillery.

While selecting a transmission site, a strategic tactic involves positioning it at a low point on a hill, behind the intended transmission azimuth. This leverages the terrain as both a reflector and an RF shade to minimize Direction Finding (DF). During our RTO Course in Wyoming, our class was strategically located near the base of a major mountain ridge, serving as a backstop for our signal. Employing Yagi antennas from our primary location to the planned site, we achieved successful voice communication and a burst data transmission spanning over 35 miles on a mere 4 watts. This scenario is akin to aiming two flashlights at each other in a dark room using light waves as an analogy.

Staying low in the terrain disperses the signal to those with higher line of sight, including aircraft. In Utah, during point-to-point communication in low valleys, the signals collection team faced extreme difficulties even when stationed atop an overlooking mountain. Despite using sophisticated signal interception equipment like a high-end Software Defined Receiver (SOR) and a tuned antenna, the scattering effect of the terrain alongside directional transmission made reception nearly impossible for them. Remarkably, the communications team encountered no such issues.

It's important to note that equipment utilized by a government-backed occupation force generally lacks additional capabilities beyond common, off-the-shelf equipment. In many cases, it's on par with or even inferior to civilian-grade technology. This insight is not intended to instill complacency; rather, it underscores the practicality of these techniques based on physics. When employing these techniques, an armed force can be well-equipped and highly proficient using inexpensive equipment.

DIGITAL ENCODING AND ENCRYPTION IN RADIO COMMUNICATION

Baofeng radio, transmitting analog signals, lacks inherent encoding or encryption capabilities. However, this doesn't imply impossibility. Despite advancements in radio technology and communication complexity, the significance of analog radio and traditional, reliable message encryption methods often get overlooked.

Distinguishing Between Encoding and Encryption

Encoding and encryption aren't interchangeable. Encoding transforms a message into a code, providing security to anyone without the means to recognize or decode it. Encryption deliberately obscures the contents of a message to conceal them. While different, when combined, they establish robust communications security. Encryption exists in two primary forms: Digital and Physical.

Digital Encryption involves software within the device to encode, encrypt, decode, and decrypt transmissions in a specific protocol. Examples include VINSON and ANDVT in US Military radios and AES encryption in DMR. Physical

Encryption, on the other hand, involves manually encrypting messages. This method is highly secure and adaptable across various transmission modes, including analog voice. For Guerrilla communication coordination, physical message encryption holds paramount importance.

The Ongoing Dynamics Recent lessons in the Ukraine-Russian War highlight these principles. Much of the communication gear used comprises commercial/off-the-shelf equipment, including both Digital Mobile Radio (DMR) and analog Baofeng radios. Initially, DMR exhibited advantages due to its encryption capability and short SMS messaging. However, countermeasures emerged; the Russians restricted their side to analog transmissions, making DMR signals distinct to the Ukrainian side. Once DMR signals became identifiable — unique in sound and appearance on a monitoring spectrum — they became prime targets for artillery and rocket fire. Consequently, both sides reverted to analog transmissions.

This adaptive strategy underscores the temporality of techniques — recognizable patterns lead to countermeasures. Intelligence relies on pattern recognition for predictive analysis. Notably, encryption is prohibited in Amateur Radio, except for specialized configurations like using a Baofeng radio with a tablet and the andFLmsg freeware app. It's crucial to note that the discussed techniques aren't intended for Ham radio use but do offer robust communication security.

Communications Security Considerations Across Categories

Sustainment-Level Communications:

Sustainment-level communications carry a relatively lower COMSEC requirement, but dismissing security concerns entirely is unwise. Even in routine or disaster communications, interception risks exist from malicious entities. Transmitting data swiftly is crucial, hence the adoption of digital communication modes like andFLmsg by some volunteer relief organizations due to their efficiency in handling larger data quantities within tight timeframes.

Tactical-Level Communications

Tactical communications demand significantly higher COMSEC measures. Analog radio usage mandates minimal power for communication within the essential distance, as discussed in the preceding chapter. However, due to the urgency inherent in tactical communication, there's often inadequate time for employing encryption methods outlined in this chapter. Modern tactical radio systems focus on real-time encryption, creating identifiable patterns and requiring specialized equipment. Given the constraints, guerrilla bands rely primarily on equipment like the Baofeng radio. The utilization of trigrams, later detailed in this chapter, serves as an effective albeit older method to obscure messages and shorten transmission duration.

Clandestine-Level Communications:

By nature, clandestine communications necessitate the highest level of COMSEC. These communications involve encrypting messages before transmission, either through analog or digital bursts. The encryption methods detailed in this chapter, particularly Trigram and One-Time Pad (OTP) cipher, offer robust encryption across various mediums. However, it's essential to note that while Trigrams and especially OTP provide strong encryption, they attract significant attention from counterintelligence agencies.

Agencies may dedicate substantial resources to pinpoint the source of such transmissions despite their encryption strength.

Digital Communications With A Baofeng Radio

To employ digital communications with a Baofeng radio interfaced with a tablet or mobile device, you can use a free app named andFLmsg. This setup allows keyboard-to-keyboard digital data bursts between two terminals swiftly and efficiently, reducing transmission time significantly and minimizing the chances of human error. While there's a more comprehensive version of the app, Fldigi, available for laptop computers, the andFLmsg app is simpler, user-friendly, and easier to troubleshoot, and this guide focuses specifically on its usage.

Tools Required:

- Baofeng APRS cable, connecting the audio jack of the tablet to the two-prong microphone jack of the radio.

Setting up the Baofeng Radio for Transmission:

1. Switch the Radio to VFO mode, displaying both the Transmit and Receive frequencies.

2. Activate TDR (MENU #7) to ON, enabling the radio and software to receive on both frequencies simultaneously.

3. Adjust the radio volume to around 2/3 of the maximum to prevent distortion of received signals in the tablet's software.

4. Maximize the tablet volume.

5. Set SQUELCH (MENU #0) to 1.

6. Enable VOX (MENU #4) at level 1, allowing the tablet's audio to transmit over the radio. The transmitter control occurs through the software.

7. Disable the Time Out Timer (MENU #9) by setting it to OFF.

Utilizing andFLmsg

andFLmsg is a free digital data protocol program offering eight overarching protocols, each with multiple sub-modes based on bandwidth, totaling 108 different communications modes. This app is designed with simplicity in mind, featuring three distinct screens that can be accessed by swiping on the tablet screen. The initial screen functions as a terminal where quick messages can be entered.

When you launch the program, you'll see the operating mode displayed at the top. In the example shown, it's set to MT63 2000 LG, indicating the broader bandwidth version of MT63 with a longer interleave. This setup sends slightly longer transmissions but includes more redundant data for forward error correction, making it a preferred mode in the field. Once your report is ready, simply tap "SEND TEXT," and the program will begin transmitting the digital encoding. During transmission, the mode indicator at the top will change from blue to gold and revert to blue once completed.

You can modify the operating mode in two ways: either by accessing the settings through the three circles in the top right corner, navigating to "SETTINGS," selecting "CUSTOM LIST OF MODES," and choosing the desired operating modes, or by swiping to the second screen, the receiving terminal, and using the "NEXT MODE" or "PREVIOUS MODE" buttons.

After sending your message, swipe to the second screen, where received messages will appear. Here, you'll notice a message received in MT63 2000 LG format, with the message header indicating the traffic nature (SALUTE) and the message content clearly displayed. The entire transmission typically takes around 3 seconds, significantly quicker than conveying the same message via voice.

The bottom buttons offer control over mode navigation and squelch levels. Adjust the squelch only as high as necessary. Monitoring the S2N (signal to noise) bar above the buttons provides a visual representation of the squelch level and incoming traffic. Insufficient squelch results in digital static with random characters, while excessively high squelch might cause message misses.

andFLmsg also supports sending images, files, and dedicated report forms. These can be imported/exported as .html files, simplifying the creation of custom report formats that can be filled in as required.

To send pictures:

1. Swipe to the third screen and tap "COMPOSE."

2. Scroll down to "PICTURE" and long-press to open the blank picture form.

3. Use "ATTACH PICTURE" to add images from the device's memory.

4. After completing the form, tap "SAVE TO OUTBOX" at the bottom left and then "RETURN" at the top to go back.

5. Access the outbox to find your message and long-press to open it for further actions.

To send the message, first, take note of the modes displayed—MT63 2000 LG and MFSK64—along with the estimated time for message transmission. MFSK64 is a high-speed, wideband mode suitable for transmitting images and is the selected setting for this operation due to its rapidity.

While other modes can send images, they operate much slower in comparison. When ready to transmit, simply tap "TX OVER RADIO," and the message will begin its transmission.

To access received message forms, return to the terminal and tap "INBOX." Here, you'll find the incoming messages, and as before, long-press to view their contents.

Trigram Encryption

Trigram Encryption is a method used to both condense and encrypt a message, substituting three letters for a word. A sample list of these trigrams can be found in Appendix C and serves as an example in this manual. While Amateur Radio operators use Q codes to shorten messages without encryption, the use of trigrams in clandestine communications involves randomizing them, requiring the current key to decrypt the message. Intercepted trigrams without the current key appear as random characters.

During the Cold War, covert teams across different continents extensively employed trigrams, transmitting Morse Code letter groups through tape recorders at high speeds to burst the transmission over analog radios. Lawrence Myers detailed this technique in his book "SpyComm," citing its use for coordinating clandestine cells in South American countries while working for the CIA. Today, a similar method can be employed using andFLmsg with a Baofeng radio.

Creating a Trigram List

This involves entering the entire list into a spreadsheet on an air-gapped computer. Organize it into four columns: the first three columns for single letters and the fourth column containing the corresponding words from the master list. Configure the columns to shift at different intervals, creating a robust encryption method that requires identification by a cryptologist and extensive data processing to break. While it's possible to decrypt, the time-consuming nature of deciphering such messages renders them practically useless by the time they are decoded, assuming interception and recognition of the digital mode used for transmission.

Using Trigrams For Message Encoding

This involves a straightforward word substitution. The list contains organized ENCODE and DECODE sections: the ENCODE list has alphabetized words, while the DECODE list has alphabetized trigrams. With practice, both functions become quick and easy to use.

Consider a sample message reporting an ambush by concealed paratrooper infantry resulting in fighter casualties in the area. The original message might be: "FIGHTERS KILLED IN ACT APPROACHING CAMOUFLAGED PARATROOPER INFANTRY." Encrypting this using our sample list looks like this: "FIGHTERS (BQL) KILLED IN ACT (WIO) APPROACHING (BKF) CAMOUFLAGED (JNU) PARATROOPER (OAM) INFANTRY (BLT)."

To maintain message simplicity, it's advisable to conclude the message with "ZZZ" to signify the end and prevent any confusion. The encrypted message then becomes: "BQL WIO BKF JNU OAM BLT ZZZ."

Transmitting this encrypted message via a data burst drastically reduces transmission time to just a couple of seconds. While there remains a possibility of interception, utilizing the methods for transmission described in this manual significantly decreases this likelihood.

One Time Pad (OTP)

Regarding One Time Pad (OTP) encryption, it's also known as a Vernam Cipher and is theoretically unbreakable when used only once. OTP employs a codex to assign numerical values to each alphabet letter. Subsequently, it utilizes a random series of numbers organized into groups of five digits each, aligning them into a block for encryption.

A-1	B-70	P-80	FIG-90
E-2	C-71	Q-81	(.)-91
I-3	D-72	R-82	(:)-92
N-4	F-73	S-83	(')-93
O-5	G-74	U-84	(,)-94
T-6	H-75	V-85	(+)-95
	J-76	W-86	(-)-96
	K-77	X-87	(=)-97
	L-78	Y-88	(?)-98
	M-79	Z-89	BRK-99
			SPC-0

32244

32244 52687 97412 86319 11011 59341 73741 29248 65123 56878

19652 15821 01512 44462 36799 12491 08122 13421 79512 72121

06503 41072 61772 22229 13714 24763 19975 59821 63483 64465

14324 86922 24922 62231 52134 70252 14319 23219 92012 08194

11119 21919 67421 01254 25722 42491 16132 71151 86117 01825

69138 19415 21792 46102 29541 17221 16413 65714 91978 62353

42551 58181 59165 14216 21238 87777 42524 61411 92206 11872

22129 20919 51252 20934 85440 22922 21922 06210 99213 32819

51261 86248 15863 14618 52148 91232 24524 14388 71341 92230

The OTP example presented here includes the codex and a sample key. Typically, the codex remains constant across all messages, while the OTP key is used only once. The key, identified by the numbers "32244" at the top and repeated as the first five-digit block, is duplicated—one for the sender and one for the receiver. This key serves as an identifier, instructing the recipient which key to use for decrypting the message and isn't encrypted using letters.

The initial encryption step involves assigning numerical values from the codex to individual letters in the message. For instance, the sample message "Meet at the old church at five PM Tues" would be encoded numerically as:

M e e t _ a t _ t h e _ o l d _ c h u r c h _ a t _ f i v e _ p m _ T u e s

79 2 2 6 0 1 6 0 6 75 2 0 5 78 72 0 71 75 85 82 71 75 0 1 6 0 73 3 85 2 0 80 79 0 6 84 2 83 0

These numerical values are then divided into five-character groups. The spacing itself is inconsequential but is done for simplicity in mathematical calculations. OTP encryption and decryption rely on straightforward arithmetic:

The process for encryption involves subtracting the numbers in the key from the encoded message, digit by digit. This difference becomes the transmitted message. To decrypt, the received message is placed above the key, and one digit at a time is added. This returns us to our codex numbers, which are then decoded into letters.

The given example illustrates the message encoded via the codex on Line 1, the OTP key on Line 2, and the encrypted message on Line 3: Line 1: - - - - - 79226 01606 75205 78720 71758 58271 75016 07338 52080 79068 42830 Line 2: 32244 52687 97412 86319 11011 59341 73741 29248 65123 56878 19652 15821 Line 3: 32244 27649 14294 99996 67719 22417 85530 56878 42215 06212 60416 37019

Note: If the bottom number is greater than the top number, add 10 to the top number. When decrypting, if the sum is greater than 10, drop the 1, leaving the second digit remaining.

Decrypting involves aligning the transmitted message with the key underneath and adding a single digit at a time. This process yields the numbers from the codex, which are then used to decode the individual letters.

Once decrypted, the message becomes: 79 2 2 6 0 1 6 0 6 75 2 0 5 78 72 0 71 75 85 82 71 75 0 1 6 0 73 3 85 2 0 80 79 0 6 84 2 83 0 M e e t a t t h e o l d c h u r c h a t f i v e _ p m _ t u e s

It's crucial to double-check the encryption and decryption of messages before sending to ensure accuracy and avoid confusion. Grid paper can be helpful in keeping the digits aligned, as the most common mistakes in OTP encryption and decryption occur due to misaligning number columns. Keeping the process systematic and working digit by digit helps maintain accuracy.

Utilizing one-time pads (OTPs) raises significant concerns regarding COMSEC (Communications Security) and Operational Security (OPSEC). It's crucial to note that for security reasons, the remainder of a pad, even if unused, should never be reused. The strength of an OTP lies in its randomness and its use only once. Creating OTPs can be accomplished through various methods, such as using dice to generate random numbers or utilizing specialized OTP printers like an ADL-1.

OTP encryption, when used just once, is considered mathematically unbreakable and offers the highest level of security possible. However, maintaining brevity in communications is essential. A recommended approach involves combining two encryption methods — Trigrams and OTP — to create a condensed and unbreakable burst. This approach may demand time and expertise but presents a highly secure transmission method. Moreover, employing dual layers of encryption serves as an advantage, especially for underground organizations, as even if one key is compromised, the other might remain secure. For a guerrilla force operating clandestinely, the utmost caution must be exercised regarding COMSEC. By employing data bursts alongside layered encryption methods like Trigrams and OTP, the most robust form of COMSEC over the air can be established.

BONUS SECTION

HOW TO DESIGN THE BAOFEG REPEATER

This tactical/portable repeater design enables UHF/VHF cross band repeater functionality through affordable and easily obtainable COTS (commercial off-the-shelf) products. Its components, including sealed boxes, solar panels, and a battery, offer the potential for prolonged operation, possibly spanning several years depending on usage. While adaptable to various radios, this design specifically incorporates Baofeng UV-SR HT radios. Additionally, an optional enhancement involves acquiring an 8-watt UV-5R8W HT for the transmit (Tx) side of the repeater pair. A schematic overview of the design resembles the following illustration:

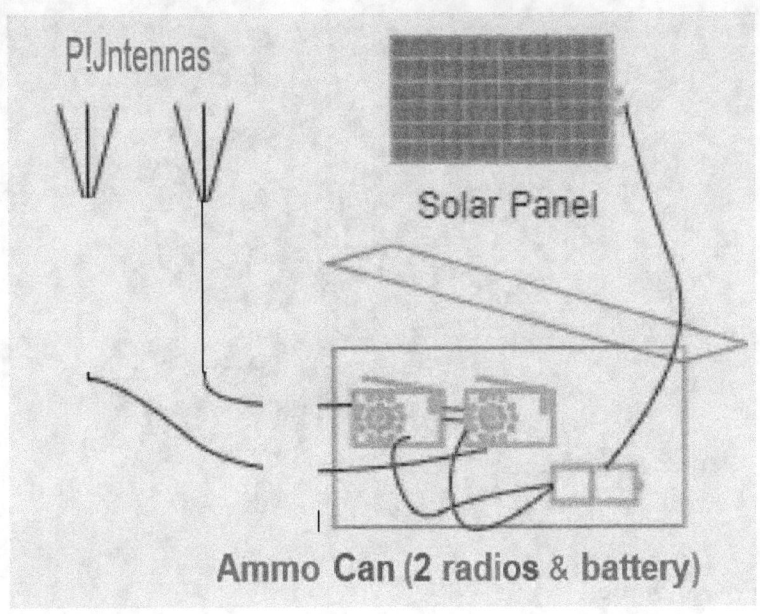

P!Jntennas

Solar Panel

Ammo Can (2 radios & battery)

Ammo Can Setup

Parts List

• 30 Cal ammo can

A typical 30 Cal ammo can is the perfect size for this project. To allow for antenna and solar panel cables, a hole is drilled into the ammo can. Use a 1" clamp connector (obtained from any hardware store) to protect and secure the wires. Silicone can be applied to waterproof the box.

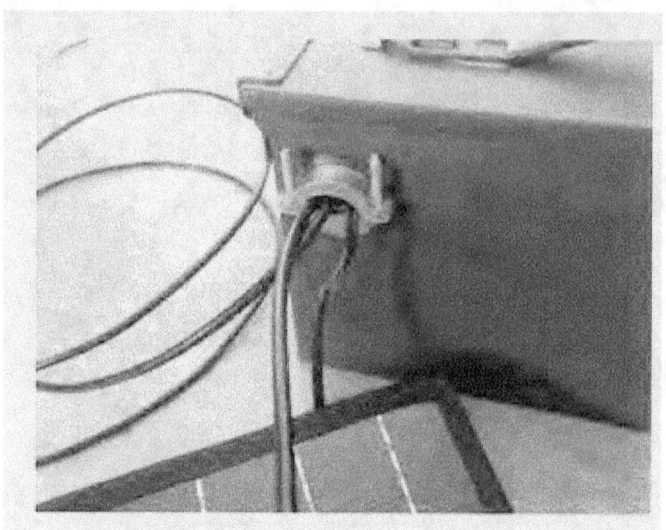

Power Setup

(Note: This part is optional but is an important part for extended operation.)

Parts List

- Cigarette 'Y' adapter
- 10-12awg Female Spade Crimp
- Standard 7.4 Ah lead acid battery
- SAE extension cable
- 5-Watt Solar Panel
- 2 X UV-SR Car charger adapter

Leaving enough wire to easily work with, cut off the male end of the 'Y' adapter and cut the SAE cable in half. Strip both wires on the end of the 'Y' adapter and the SAE cable. Twist the red wires together and attach the spade adapter. Repeat this process for the two black wires.

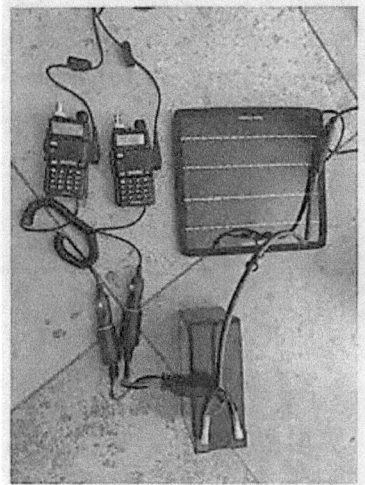 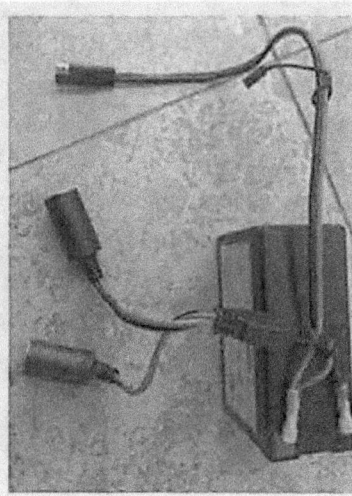

If simplicity is your priority, you might consider skipping the ammo can, battery, and solar panel setup and opt for a more straightforward design, such as this:

THE REPEATER SETUP

Parts List:

- 2 X Baofeng UV-SR

- Baofeng UV-5R8W

- Repeater box (Note: The 'YiNi Tone RC-108' has worked well, but it might be unavailable on Amazon. Try checking Walmart.com or eBay.com for the 'RC-108 Repeater Box'. Other options exist on Amazon, although they haven't been tested yet.)

This guide includes programming instructions for these radios using both the face controls and screen shots from Chirp software for those preferring a programming cable. Check Appendix B for recommended general CHIRP settings applicable to all radios.

Before programming the repeater pair, ascertain three key details:

- Transmit (Tx) frequency. (For instance, we'll use 146 MHz in this example)

- Receive (Rx) frequency. (For instance, we'll use 447 MHz in this example)

- CTCS Tone. (For instance, we'll use 67 Hz in this example) Note: While not a security measure, setting an R-CTCS tone on the Rx side of the repeater pair can deter unauthorized users from triggering the repeater.

Note :Any frequency that your radios support can be used by this repeater

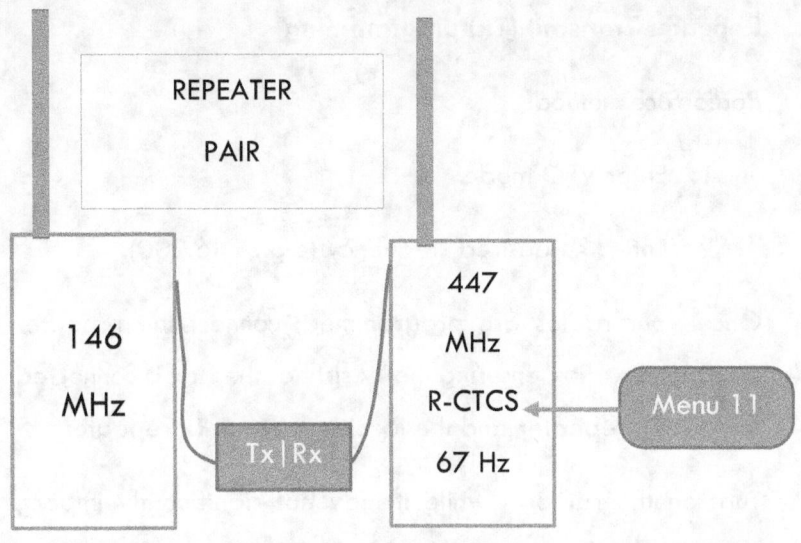

Repeater Receive (Rx) programming

Radio face method:

1. Enter VFO mode.

2. Enter the desired frequency (e.g., 447.000).

3. Press [Menu] 11 [Menu] 67 [Menu][Exit].

   - (Note: Alternatively, use the up/down arrows to select the desired tone frequency instead of entering '67'.)

Repeater Transmit (Tx) programming

Radio face method:

1. Enter VFO mode.

2. Enter the desired frequency (e.g., 146.000).

Once your radios are programmed, connect the repeater box to the radios, ensuring the Tx side of the box is connected to your Tx repeater and the Rx side to your Rx repeater.

Turn on the radios - while it may not significantly impact, setting the volume between 50% to 75% has shown positive results in my experience.

Team Radio Setup

Before programming the team radios, determining the offset is essential. To find this, subtract the Rx frequency from the Tx frequency. If the result is positive, it's a positive offset; if negative, it's a negative offset. For instance, subtracting 146 from 447 gives us a positive offset of 301.

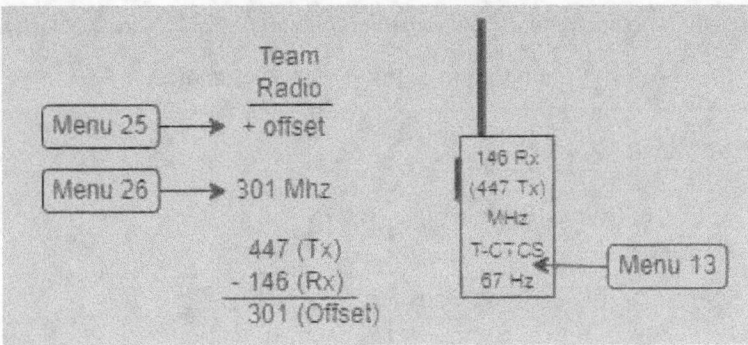

Radio face method:

1. Enter VFO mode.

2. Input the desired frequency (e.g., 146.000 - ensure the correct offset calculation).

3. Press [Menu] 25 [Menu] <use up/down arrows until '+' is displayed> [Menu] [Exit].

4. Press [Menu] 26 [Menu] 301000 (for a positive offset of 301) [Menu] [Exit].

5. Press [Menu] 13 [Menu] 67 (or select the desired tone frequency) [Menu] [Exit].

The optional antenna build requires an external antenna, particularly if you intend to use an ammo box to house your repeater. If you prefer not to construct antennas, there are suitable options available for purchase on platforms like eBay. One recommended type for this application is the 'Dual band Slim Jim' or the Jungle Antenna. Ensure the antenna's cable end matches your radio—utilizing a BNC adapter for antennas with a BNC end is my preferred choice.

Parts List:

- SMA-F to BNC-F adapter (2 pack)

- 20' RG-58 coax cable

- Heat shrink tubing (various sizes)

- BNC-M crimp connector

When assembling this antenna, precise measurement is crucial. In initial attempts, slight discrepancies (3-5mm) resulted in high SWR. Improved accuracy lowered the SWR to around 1.2 at 145 MHz. You'll need tools like a soldering iron, wire cutters, coax strippers, a box cutter or razor blade, and a crimper for this project. To prevent movement, I used heat shrink tubing on the outer insulators and applied heat shrink to create a rope loop at the antenna's end for attachment. Additionally, I soldered approximately 5 feet of antenna feed line directly to the antenna and crimped a BNC-M connector at the end.

www.ingramcontent.com/pod-product-compliance
Lightning Source LLC
Chambersburg PA
CBHW062324290526
45794CB00005B/1888